Injecting Creative Thinking into Healthcare

The book sets out to inform a broad range of professionals working in medicine and healthcare about how creative thinking and design concepts can be used to innovate in providing an enhanced patient experience. It outlines these concepts as a primary means to identify, clarify and resolve some of the process improvement and enhancement challenges in healthcare delivery. It demonstrates by example how such challenges can be addressed, drawing on case examples from healthcare and other industries, and from the authors' own experiences as innovators and educators. It emphasises the value of learning in action. For the reader who already has a leaning towards novel approaches to addressing healthcare delivery challenges, it provides guidance on harnessing team inputs and engaging with a network of contributors. It is an ideal resource for all working in medicine and healthcare, from managers, nurses, doctors, administrators, executives and allied health professionals to medical engineers, medical physicists, medical scientists and medical product developers.

Features

- Provides a unique framework to conceptualise innovation in healthcare and medicine.
- Authored by an award-winning medical scientist and an established business school professor who have proven track-records with innovation in education settings and as entrepreneurs.
- Presents a clear interdisciplinary approach, complemented with practical case studies set in the context of the challenges facing healthcare delivery in the 21st century.

Barry P. McMahon has a national and international reputation as an academic medical physicist in the fields of novel physiological measurement and medical device innovation and design. He is the co-inventor of the Functional Lumen Imaging Probe (FLIP) technique later commercialised as EndoFLIP™. He was the Director of the Innovation Academy at Trinity College Dublin from 2012 to 2017. Since 2020 he is advising Children's Health Ireland on innovation practice. In 2019, he retired as Chief Physicist/Clinical Engineer at Tallaght Hospital, Ireland, and currently runs his own innovation-consulting group, Electric Mindset Ltd.

Paul Coughlan is Professor in Operations Management and Co-Director of Faculty at Trinity Business School, Trinity College Dublin. His research explores the collaborative strategic improvement of operations through network action learning. He was the Director of the Innovation Academy at Trinity College Dublin from 2010 to 2012. He is a founding director of a research-based spin-out venture, Easy Hydro Ltd.

Injecting Creative Thinking into Healthcare

Bringing Innovative Practice to Healthcare

Barry P. McMahon and Paul Coughlan

CRC Press is an imprint of the
Taylor & Francis Group, an **informa** business

First edition published 2023
by CRC Press
4 Park Square, Milton Park, Abingdon, Oxon, OX14 4RN

and by CRC Press
6000 Broken Sound Parkway NW, Suite 300, Boca Raton, FL 33487-2742

CRC Press is an imprint of Informa UK Limited

British Library Cataloguing-in-Publication Data
A catalogue record for this book is available from the British Library

Library of Congress Cataloging-in-Publication Data
Names: McMahon, Barry P., author. | Coughlan, Paul (Professor of operations management), author.
Title: Injecting creative thinking into healthcare : bringing innovative practice to healthcare /
Barry P. McMahon and Paul Coughlan.
Description: First edition. | Boca Raton : CRC Press, 2023. | Identifiers: LCCN 2022060054 |
ISBN 9780367641122 (hardback) | ISBN 9780367643157 (paperback) | ISBN 9781003123910 (ebook)
Subjects: LCSH: Health services administration–Ireland–Case studies. |
Health facilities–Personnel management–Case studies. | Medical innovations–Ireland–Case studies. |
Creative thinking–Ireland–Case studies.
Classification: LCC RA971 .M4338 2023 | DDC 362.109415–dc23/eng/20230419
LC record available at https://lccn.loc.gov/2022060054

ISBN: 9780367641122 (hbk)
ISBN: 9780367643157 (pbk)
ISBN: 9781003123910 (ebk)

DOI: 10.1201/9781003123910

Typeset in Minion
by Newgen Publishing UK

Contents

Acknowledgements

THE BOOK IS BASED on our shared experience of developing and delivering educational initiatives in the area of creativity and innovation, most recently in healthcare in Ireland. The roots of our approach are in practice and in a novel initiative set in Trinity College Dublin which aimed to generate awareness among doctoral students of the innovation potential in their dissertations.

We acknowledge the following:

Trinity College Dublin for its development of the Innovation Academy concept and providing us with the opportunity to develop, lead and run this programme for doctoral students. From the start of the first programme to the publication of this book, it has taken 12 years.

Trinity College Dublin, School of Medicine, for its development of the Postgraduate Diploma in Healthcare Innovation, now in its fourth year. Our active collaboration and engagement with healthcare professionals have supported the co-development and co-delivery of our ideas.

We most of course acknowledge both our families, and in particular our ever-patient wives who supported us in many different ways through the journey this book has taken us on. We would like to thank the healthcare staff and educators we have engaged with us over the last five years and, in particular, Fran Hegarty, Ann Quinn, Marie-Claire Kennedy, Michelle Armstrong and Alison Keogh who have worked passionately and directly with us to develop innovative workshops and programmes for healthcare professionals.

Finally, CRC Press for their support in bringing our ideas together in a comprehensive book of this kind.

About the Authors

THIS BOOK IS WRITTEN by two authors who have approached creativity and innovation in healthcare from related but different perspectives.

Barry P. McMahon has a national and international reputation as an academic medical physicist in the fields of novel physiological measurement and medical device innovation and design. He is the co-inventor of the Functional Lumen Imaging Probe (FLIP) technique later commercialised as EndoFLIP™. He has won a number of prestigious awards for his research, including a Distinguished Scholar Award from the Chinese University of Hong Kong and has successfully fostered a number of commercial offshoot companies. He has published a substantial body of work in several high-impact journals related to his field of interest. Prof. McMahon was the Director of the Innovation Academy at Trinity College Dublin from 2012 to 2017. He has secured significant funding from industrial grants to support his work, and has funded several PhD and MSc projects at both Trinity College Dublin and Institute of Technology Tallaght. Prof. McMahon has established many links in Ireland and internationally and is central to several multidisciplinary projects, in particular, he has demonstrated an ability to support translational research programmes. In 2021, he retired as Chief Physicist/Clinical Engineer at Tallaght Hospital and currently runs his own innovation-consulting group, Electric Mindset Ltd. He is a former Vice-President and Governor of the Irish College of Medical Physicists and former President of the Irish Association of Physicists in Medicine. He is a co-founder of Trinity Academic Gastroenterology Group. Since 2020 he is advising Children's Health Ireland on innovation practice.

Paul Coughlan is Professor in Operations Management and Co-Director of Faculty at Trinity Business School, Trinity College Dublin. His research explores collaborative strategic improvement of operations through network action learning. His fellow researchers are active in different domains and in practice, both nationally and internationally. He has contributed actively to EU-funded research projects exploring operations improvement, innovation in food and in environmental sustainability of water distribution. He brings to this research a questioning, interventionist and reflective approach which combines action research, action learning, operations management and innovation research. He has published in various journals including *Creativity & Innovation Management*, the *British Journal of Management* and the *International Journal of Operations and Productions Management*. His co-authored book with David Coghlan, *Collaborative Strategic Improvement through Network Action Learning: The Path to Sustainability*, was published by Edward Elgar (2011). Previous academic appointments were at Aalborg University, Denmark; London Business School, UK; and University College, Cork, Ireland. He holds a PhD from the University of Western Ontario, Canada, and MBA and BE degrees from University College, Cork. He has been President of the Board of the European Institute for Advanced Studies in Management, chaired the Board of the Innovation & Product Development Management Conference and a been member of the Board of the European Operations Management Association. While at Trinity College Dublin, he has been elected to Fellowship and awarded a Provost's Teaching Award in addition to Trinity Business School Teaching and Research Excellence Awards. Prof. Coughlan was the Director of the Innovation Academy at Trinity College Dublin from 2010 to 2012. He is an Honorary Fellow of the European Operations Management Association and Honorary Member of the Continuous Innovation Network. He is a founding director of a research-based spin-out venture, Easy Hydro Ltd.

CHAPTER 1

The Opportunity and the Challenge

SO, LET'S LOOK AROUND. What is happening in the world of healthcare? Fortunately for society, professionals continue to be attracted into the area where they learn and practice the art and science of caring. Their focus is captured often in the description of specialities, many of which are novel and revolutionary, while others are incremental developments of established practice. Behind these activities we find opportunities for creative thinking and innovation. However, rarely are these skills recognised explicitly. So, it was with great interest that we spotted a recent job advertisement for a new position in the team charged with the development of a new children's hospital in Ireland. The ad is shown in Figure 1.1.

What was going on here? Who were they looking for? What might they do? And, closer to our interests, who might they work with, and what education might they require? Keep these questions in mind as we develop our thinking and introduce the book.

We begin with three questions: What is going on? Who is doing what? How can we do more? The questions are connected. The "what" question challenges the relevance of our focus on creative thinking and innovation to the opportunities and challenges in society and, in particular, maintaining and improving the health of that society. The "who" question is to help identify our collaborators, fellow travellers or independent travellers - those

DOI: 10.1201/9781003123910-1

FIGURE 1.1 Recruitment advertisement.

who have the potential and opportunity to take action, and those who can set the scene so that the proposed creative thinking and innovation can develop. And, finally, the "how" question is to scope out the educational opportunity for those with that potential.

1.1 WHAT IS GOING ON?

Many descriptors are applied to healthcare system: profession, calling, etc. Implicit in such descriptions is a sense that there is a system which delivers healthcare outcomes. However, key to that system are the people who take on the challenge of learning, qualifying, practicing and continuing to learn. The context within which this system and these people practice is one characterised by constant change, new and ever-changing treatments, budget restrictions and, often, the shock of unanticipated demands. So, in addition to a core capability in a particular aspect of healthcare, the professionals also need to be creative, resilient and innovative.

So, let's return to the advertisement shown earlier. Why does a new children's hospital need an Innovation Manager in healthcare? Is it a role where the right person can build a career? Who could fill the role? How might it be graded relative to roles and other areas of responsibility? The recruitment team in the hospital will deal with these practical questions and, hopefully, the appointee will find a way to lead, facilitate and contribute to the realisation of opportunities for innovation in the hospital. However, as the physicists remind us, to every action there is an equal and opposite reaction. Here, how might the incumbent healthcare professionals view this new role? Is it relevant? A waste of time? Nothing to do with me? These are not trivial questions as they reflect the boundaries and interconnections among the established areas of practice. However, they might also reflect a blind spot – opportunities for creative thinking and innovation permeate all of the areas of activity and can achieve greater impact if recognised, developed and implemented consciously.

Conscious recognition, development and implementation require a mindset that can overcome a natural response to a new area of activity and engagement, such as innovation management, and acceptance of different skills. There is a school of thought which sees innovation as disruptive. However, that disruption may not necessarily be destructive. Rather, it might challenge, for example, pre-existing procedures. For those responsible for implementation of these procedures, such a challenge may extend beyond the substantive content of the procedure and intrude into their perceived status and standing in the organisation. However, these realities need not stop the creative thinking and innovation. Instead, they offer a challenge back to pitch and defend the relevance of the innovation in prospect. And those innovators responsible for that pitch may need to think like entrepreneurs, if extending beyond their areas of responsibility, or intrapreneurs, if challenging within their areas. Regardless, like the pirates that Sam Conniff (2018) writes about, they need to shift their mindset, be willing to think differently, to challenge and be challenged, and to stop asking for permission to do what they know is right.

1.1.1 A Healthy Society

There is a saying, "well, at least you have your health". It is timeless and, yet the expectation of what it is "to have your health" is changing continuously. Is it timely access to treatment, prevention or mitigation? Is access a right or a privilege? Is it free or costly? Further, there remain significant differences between such expectations in the developed and developing worlds and, even in the developed world, between those with more or less wealth. Yet, all rely on the healthcare system and citizens continue to expect more for less. Such expectations translate into challenges and opportunities for innovation in technologies, procedures, processes and organisations. And, if these innovations translate into an improved system, then society as a whole benefits while individual citizens enjoy the prospect of a better life.

1.1.2 The Needs of Patients

At some stage in our lives, we all are patients. When born, many experience the healthcare system. As we grow, preventative medicines and immunisation initiatives try to ensure that we do not become patients. As we live and age, we are more likely to find ourselves engaging with the healthcare system as patients. So, what is it that patients need? Most wish to be well again and with their quality of life restored. When engaging with the healthcare system they need timely access, appropriate treatment, recovery and respect, ideally without consequential financial worries. While the organisation of healthcare delivery may group patients by commonality of medical conditions, there remains the individuality of the patients, each with their personal medical and family histories. This characterisation of patients rules out a "one-size-fits-all" approach to care which is broadly recognised by the healthcare system. However, workloads and pressure for positive outcomes can mean we are forced to exploit economies of scale and scope to deliver a basic level of care. This has been demonstrated quite specifically during 2020–2021 with the dramatic onset worldwide of the Covid-19 viral pandemic. Consequently, innovations that might impact the patient experience need to consider both the personalised treatment of each patient and the patient as one among many in, what is becoming, an increasingly complex system.

1.1.3 The Skills of Healthcare Professionals

Healthcare is a highly regulated area of activity. Education and training prepare the individuals for safe and effective practice. Lifelong learning has particular meaning in this area as the rate of change in society, techniques and technologies offer new hope for better quality of life. One way to characterise the skills of healthcare professionals is to distinguish between core discipline-based skills and broadly defined relational skills. Many discipline-based skills are acquired in third-level education programmes of study and refined in supervised practice settings. They centre around analytic thinking and problem-solving within these chosen professions. In contrast, many relational skills are built up over life but enhanced in the particular context of healthcare delivery. Here, these skills often centre around the ability to apply knowledge in teams when dealing with problems.

Creative thinking and innovation are at variance with the highly regulated and controlled area of healthcare practice. However, concern for the professional execution of appropriate quality of diagnosis, treatment, monitoring and care of patients does not mean that there is no room for improvement. Opportunities and challenges do exist and particularly in the healthcare delivery process. It is here that opportunities and challenges emerge to deliver better quality, faster, more dependably, flexibly and at an affordable cost. An injection of creative thinking and innovation into healthcare is to explore and exploit these opportunities and challenges. To play their part, the highly educated and trained professionals need to be empathisers, communicators and problem-solvers in the areas of their healthcare operations and medical practice.

1.1.4 The Enabling Context of Delivery

Healthcare delivery takes place in many and diverse contexts or settings. The setting may be one where many activities are undertaken for and by many people simultaneously. Or it may be one where there is a specialised and customised interaction between professional and patient. It could also be one in which the task requires different degrees of interaction among the professionals and with the patient. The setting may change as day turns to night, as the seasons change or as the criticality of the case emerges and develops. Finally, delivery may be of a routine procedure for which all know their roles. Alternatively, it may be so novel as to make planning the steps or anticipating the outcome difficult.

In such a challenging and often life-threatening context, the pressure of the moment may rightly dominate and reduce the opportunity for creative thinking and innovation. Or, indeed, the opposite might be the case – the routine and standard procedure may not be the pathway to addressing the patient's condition. Yet, it is here that all healthcare professionals practice and develop their roles. A key proposition underlying this book is that the role of the healthcare professionals, be they clinicians, nurses, allied health professionals or administrators, needs to include responsibility for continuous process improvement. In a process, a series of tasks combines to make something (hopefully desirable) happen. Information, materials and people flow within and between the tasks. The tasks may generate or consume materials and information which need to be stored securely and, yet accessibly. Improvement opportunities may occur at critical control points and require those responsible for the process to be able to look at what is going on in order to see, for example, the opportunity within traditional practice to innovate toward improvement. Such opportunities may present themselves in terms of digitisation – the replacement of discrete processes or procedures with digital analogues – or digitalisation – the use of digital information to revisit and change decision-making processes and architectures. Either way, modern creative concepts such as creative thinking, design thinking, process innovation and evaluation may be able to help. However, it is here that many healthcare professionals may feel that this challenge is beyond their role, their responsibility and indeed, their skill set. Our contention is that this is simply not true and that without their active engagement in a method to better understand the challenge, be creative about ideas that might solve it and be active in the formation and delivery of the desired improvements, this work will never be realised or at least not optimally so.

1.2 WHO IS DOING WHAT?

The scale and scope of creative thinking and innovation in healthcare varies but the possibilities are endless. It is often principally related to people, systems or processes, or technology. It can occur at any stage of the caregiving continuum from the hospital to the community, from the bedside to the patient's home.

So, let's focus first on an innovation in which one of the authors has been directly involved. In Chapter 2, we will tell this story in detail. In brief, as a medical physicist, the author co-developed a new technology known as EndoFlip® (endolumenal functional lumen imaging probe). It is a new, minimally invasive device created to complement traditional diagnostic tests. It uses a balloon mounted on a thin catheter placed trans-orally at the time of a sedated endoscopy. It offers the capability of measuring the cross-sectional area and intraluminal pressure of the oesophagus while under distension (as if a solid bolus was present). An early step was to outline the need in practice and the shape of a solution. Over a number of years, the medical physicist collaborated with a medical practitioner as they worked towards that solution. The device is now used widely in practice.

In the second short example, a community nurse observed that most wound dressings were extremely expensive. Choosing the correct dressing for the patient can reduce the number of dressings required and reduce the overall costs to the health service. In response, she developed a novel teaching platform in consultation with her community nurse colleagues. Her objective was to provide information about the various wound dressings available, and education on the different types of wounds the nurses would come across during their work in the community.

So, as a former US President stated and urged, "Yes, we can". There are many opportunities for change and improvement. It does not take a specialist or genius innovator to identify and develop them. Our contention is that all healthcare professionals have the opportunity and capability to question, to take action and to reflect on the outcomes.

1.3 HOW COULD WE DO MORE?

We have our objectives for this book. However, on its own, the impact of the book will be limited. The context for creativity and innovation in practice is crucial. It needs to build recognition and respect for innovation and, crucially, enable continuing education and training. A novel programme of study began in Ireland in 2019, titled the Postgraduate Diploma in Healthcare Innovation at Trinity College Dublin. Spanning a full year and bringing together a range of healthcare professionals, the objective has been to develop the capability and confidence of participants to lead innovation, to apply new technologies in practice and to implement solutions.

The participants are drawn from a wide spectrum of specialisms: general practice, midwifery, community practice, pharmacy, physiotherapy, surgery, to name but a few. The contributing faculty are drawn from healthcare practice, health economics, design thinking, organisational design and operations management. Together, the participants develop competence in a new language – that of creative thinking and innovation. They come to recognise ideas with potential and the value of collaboration in realising the potential of those ideas. The collaborating voices may come from within the participant innovator's team or elsewhere in the healthcare delivery system. The voices may come from ranks above or below that of the participant. The mindset shift and cultural supports required are not trivial. However, the programme is an example of an initiative with real potential.

1.4 THE PLAN FOR THE BOOK

It would be easy to describe the book as a piece in three parts. However, such simplicity is to disguise the evolutionary nature of the structure and the balance of the content. The roots of the book are in practice, and in a philosophy that values intervention in that practice. So, while research based, it is not an academic treatise on creative thinking and innovation. Rather, it is presented as an accessible, easily read exploration of actionable practice to support our ambition for the reader – "yes we can".

Organised in eight chapters, we have grouped them into three parts:

- The opportunity to make a difference for people.
  - This part deals with the situation, the opportunity and the underlying values.
- The underlying principles to guide our thinking and action.
  - What are the foundations that help to set our direction?
- The process to achieve and to learn.
  - What process, tools and techniques can support creativity in practice?

The book is not a novel. It is not a work of fiction where, as a reader, you try to avoid looking at the last page until you arrive there. No, after reading this first chapter, we invite you to read the final chapter for the first time. It describes the actual implementation of the ideas developed in the book in a practical healthcare setting. Don't worry if some of the terms, techniques or language are foreign to you. They will not remain so after you have read the other chapters in sequence and return again to the last chapter.

Enjoy!

SOURCES AND SUGGESTED FURTHER READING

Conniff, S. 2018. *Be more pirate*, New York, Ataria Simon & Schuster.

CHAPTER 2

Understanding Creative Thinking, Design and Innovation

2.1 HOW WOULD YOU KNOW IT IF YOU SAW IT?

We present two case studies, one of which is in the healthcare setting. Both are real cases (rather than fictional or "arm-chair") in which the authors feature directly. In the presentation of the cases, we are ambitious to demonstrate our active engagement in the practice of injecting innovation into practice. Each case demonstrates concepts that support innovation in practice:

- Be fearless and innovate.
- Engaging in disruptive and creative thinking to address a healthcare problem.

2.2 CASE 1: BE FEARLESS AND INNOVATE

Action-oriented research focuses on studies and critical analysis that are geared towards enhancing practice and theory directly in a particular field of study. Some years ago, one of the authors was invited to deliver a keynote address to an academic conference focused on action-oriented research. The conference organisers offered 90 minutes on the programme for the address. Now, the idea of speaking for that length of time was not so much

DOI: 10.1201/9781003123910-2

daunting as of questionable value: how to hold the attention of an audience of research-active academics for a period longer than a normal lecture. From his notes:

> *I thought about it for a while and then had an idea. The focus of the conference was on action-oriented research, where researchers engaged actively with the topic and carried out the research with people rather than on people. As a group, they were well-experienced and drawn from many disciplines in the arts, humanities and social sciences. Yet, as researchers they all faced the challenge of taking the first step–making that phone call to a key informant, meeting with a group, leading a diagnostic discussion. Having been in that situation myself, I recognised the challenge and, critically, linked it to a concept in the design management literature–creative confidence. And, so, I built my keynote session around demonstrating creative confidence in action and linking it to the execution of an innovative and interventionist research design.*
>
> *It was fun–serious fun, but fun. I adapted the IDEO shopping cart/trolley challenge where a multi-disciplinary team of designers, engineers and retailers work together to reimagine the shopping trolley (see sources in 2.5). But, at the outset, the audience did not know that. Rather, I began by prompting the audience to look around and to see how the outcomes of operations processes were everywhere: everything in the room had been produced by an operation; every product/service the audience members had consumed that day (shampoo, radio, taxi, coffee, proceedings…) had been delivered by an operation; and, that Operations Management thinking underpinned the design, planning, control and implementation of everything they bought, sat on, wore, ate, threw at people and threw away. So what? Well, as action-oriented researchers, they all faced the challenge and opportunity to learn from the applied activity that characterises practice. I reminded them of their familiarity with designing an action-oriented research initiative (Coughlan & Coghlan, 2016):*
>
> - *framing the issue;*
> - *determining the scope;*
> - *gaining access;*
> - *negotiating a role;*
> - *and developing an ethical position.*

It followed then to remind them of the steps in implementing the research design: understanding the context and purpose; constructing an intervention; planning the actions; taking action; and evaluating the action. Finally, I reminded them of the skills required for such action-oriented research (Coughlan & Coghlan, 2016):

- *collaborative intervention;*
- *analysis;*
- *learning-in action;*
- *journal keeping.*

Now so far, I had not said anything that most would have regarded as new. But then, I prompted them to consider if designing and implementing such a research initiative was easy. That prompt took them into the design management domain.

So, I invited them to think as designers where a design is form imbued with meaning (…and the meaning changes). As designers:

- *they had choice in a process of discovery and creation:*
- *what meaning to choose;*
- *how to discover people's explicit and implicit needs;*
- *how to create appropriate new solutions.*

In response, as designers, they needed to think. So, I introduced them to the concept of design thinking and assured them that, just because they didn't dress in a particular way or hadn't gone to a design school, they were designers. They could enact design thinking in the design and implementation of their action-oriented research through:

- *being empathetic,*
- *integrative,*
- *optimistic,*
- *experimental*
- *collaborative.*

This was a new perspective on their pre-existing understanding of themselves as researchers. But to be effective as designers they needed to be creative.

Creativity in their workplace was not something that the audience members might have thought about. So, I reminded them that creativity was something to be practiced, not just a talent with which they were born. I asked what might stop them, leading them into identifying fears: fear of the messy unknown; fear of being judged; fear of the first step; fear of losing control. I prompted them to recognise a need for creative confidence (Kelley & Kelley, 2012). It was the natural ability to come up with new ideas and the courage to try them out; it meant having the humility to let go of ideas that were not working and to accept good ideas from other people; it required overcoming the fears that hold many back; it offered the prospect of breaking challenges down into small steps and then building confidence by succeeding on one after another.

So what, I asked? Was there an equivalent creative confidence challenge facing them as action-oriented researchers? How would they recognise it? How would they overcome it? Then, I encouraged them to step way out of their comfort zones.

"Let's give it a shot! Take a well-known object with which there are some problems (and opportunities). (Re-)discover (your) creative confidence as a shopping trolley/cart designer." With that, one of the conference organisers pushed in a shopping trolley, borrowed from a local supermarket, into the centre of the room. Uncertain and bemused, they smiled nervously as I broke them into teams and set out their mission (Figure 2.1):

What followed was silence, nervous looks, participants reluctantly rising from chairs and then…and then noise! Within minutes, there was the spectacle of a senior academic of some repute sitting in the shopping trolley while one of his team members tried to evaluate manoeuvrability. And, so it went. The participants took their specific challenges seriously. They clarified the context and purpose; constructed an intervention; planned their actions; took action; and evaluated the action. From the outset, they had to overcome their fears of the messy unknown of trolley design and dynamics, of being judged by peers, of taking the first step which might be irreversible; of losing control in a task outside of their comfort zone.

It worked! For the last half an hour of the keynote address, we reflected on the experience. We explored the context and purpose of the creative design activity. We described what had happened, listing and naming events and incidents. We considered some actions and clinical interventions still in prospect. But then, we took the reflection in a different direction: I asked about how their (re-) discovery of creative confidence might be understood in terms of exploiting their expertise as action-oriented researchers. They saw themselves in a new light.

It was a most unusual keynote address. And, in truth, it was not just about inviting a group to recognise and to demonstrate creative confidence. Rather, for the speaker, it also required a degree of such confidence. Creativity and innovation are like that, and, without confidence, nothing will progress. In effect, innovation is much better understood (and implemented) with action, which requires confidence.

2.3 CASE 2: DISRUPTIVE AND CREATIVE THINKING TO ADDRESS A HEALTHCARE PROBLEM

Many healthcare problems require new thinking which may disrupt existing processes and practices and may make existing systems redundant. However, the possibility of such disruption should not, of itself, impede the innovation. The hands-on experience of one of the authors illustrates this challenge as he engaged in the design of a new measurement method that offered clinical value.

Your mission (should you choose to accept it):

Team	Specific Challenge	Generic Challenge
1	Safety – you, others, children...	(Re-)discovering creative confidence • Process • Organisation • Culture • Management •...
2	Manoeuvrability – in store, car park...	
3	Segregation of shopping– dry, perishable, toxic...	
4	Environmental friendliness– materials, life, after-life	
5	Security – theft, vandalism, identity...	

FIGURE 2.1 Your mission.

In humans, one of the primary functions of the gut (or digestive tract) is to carry ingested food from the mouth into the body for sustenance. At its simplest, this "transport system" is constructed of a series of pumps, chambers and valves that control the flow of the ingested material. The valves play a primary role in controlling flow and stopping or reducing back flow. A great example of an important valve in this physiological process is the lower oesophageal sphincter (LOS). The LOS controls the release of food from the mouth and gullet into the stomach where it is broken down with acid. This valve also ensures that stomach acid and materials do not travel back up towards the mouth. The medical condition where acid and stomach contents do travel back is called heartburn or reflux disease. It is mostly attributed to a badly functioning valve. So, diagnosing whether the valve is working or not can be important for treatment and monitoring of the disease.

The traditional measurement method for establishing whether the LOS is working or not is known as manometry. It involves the use of a sensor or sensors on a tube or catheter that is inserted into the patient's mouth and placed across the valve to measure its tightness or squeeze function. However, in determining valve function, manometry has been shown to be an extremely basic tool with very limited functionality in determining disease states and in diagnosing or monitoring a condition. A barium oesophagram is another traditional test. The oesophagram visually demonstrates the clearance of a bolus – swallowed material – from the oesophagus. And while manometry measures the strength of contraction in the oesophagus, the probe used is very thin and may not accurately represent the mechanics of the oesophagus in response to a bolus of food.

A novel technology known as EndoFlip® (endolumenal functional lumen imaging probe) is a newer, minimally invasive device created to complement traditional diagnostic tests. EndoFlip® uses a balloon mounted on a thin catheter placed transorally at the time of a sedated endoscopy. In comparison to the traditional diagnostic tests, EndoFlip® offers the additional capability of measuring the cross-sectional area and intraluminal pressure of the oesophagus while under distension (as if a solid bolus was present).

How did this innovation of EndoFlip® take place? It was a collaboration over a number of years between one of the authors – an Irish medical physicist – and a Danish medical practitioner. Looking at the story from the perspective of the physicist is particularly telling. Through the nineties, he was curious and fascinated with new systems and tools to provide patient care. However, he always homed in on solutions, looking at what was practical and feasible. In every sense he was an innovator.

In the early years of the new millennium, the physicist opened a new line of activity. As a reflective practitioner, he began to engage creatively at the intersection between innovation and invention. The forum for this engagement was his home-university-based Innovation Academy. In particular, he was captivated by the notion that innovation was not just about technology and computers but especially about people and process. He was intrigued by descriptions of what others had done and found them to be so powerful as he reflected on his experience and plans. By then, the EndoFlip® concept had formed and people asked: How did it happen?

The EndoFlip® story is easily told but contains all of the complexity of an innovative solution to a known problem and a gateway to, then unknown, applications. Figure 2.2 illustrates the timeline from bright idea to a functioning system.

An early step was to outline the need in practice and the shape of a solution. Certain medical conditions require suturing the oesophagus, but doctors had to rely on symptomatic evidence to confirm that their suturing worked. So, the question arose: how might a probe enable a new step in the process: how might it combine with an endoscope to measure the suturing?

> *When we looked at the need for a solution to this problem, we questioned whether, instead of measuring how closed or tight the valve is, we should be measuring how open or relaxed it is. The idea was quite radical at the time and certainly disruptive. So, we figured out how to place a precision probe into the valve and measure the precise area inside the valve region (Figure 2.3).*

The medical physicist engaged actively with the medical practitioner and benefitted throughout from interactions with many others. He built a picture around the invention, developing a plan for the project. User experience, empathy, prototyping and trust were central to the plan. As an innovator, he took a chance on this idea. He noted that some doctors might be willing to collaborate. He saw the link to design thinking and went to medical practitioners who led in the area of diagnostic tests. He let them play with the prototype probes, watching and listening to their feedback.

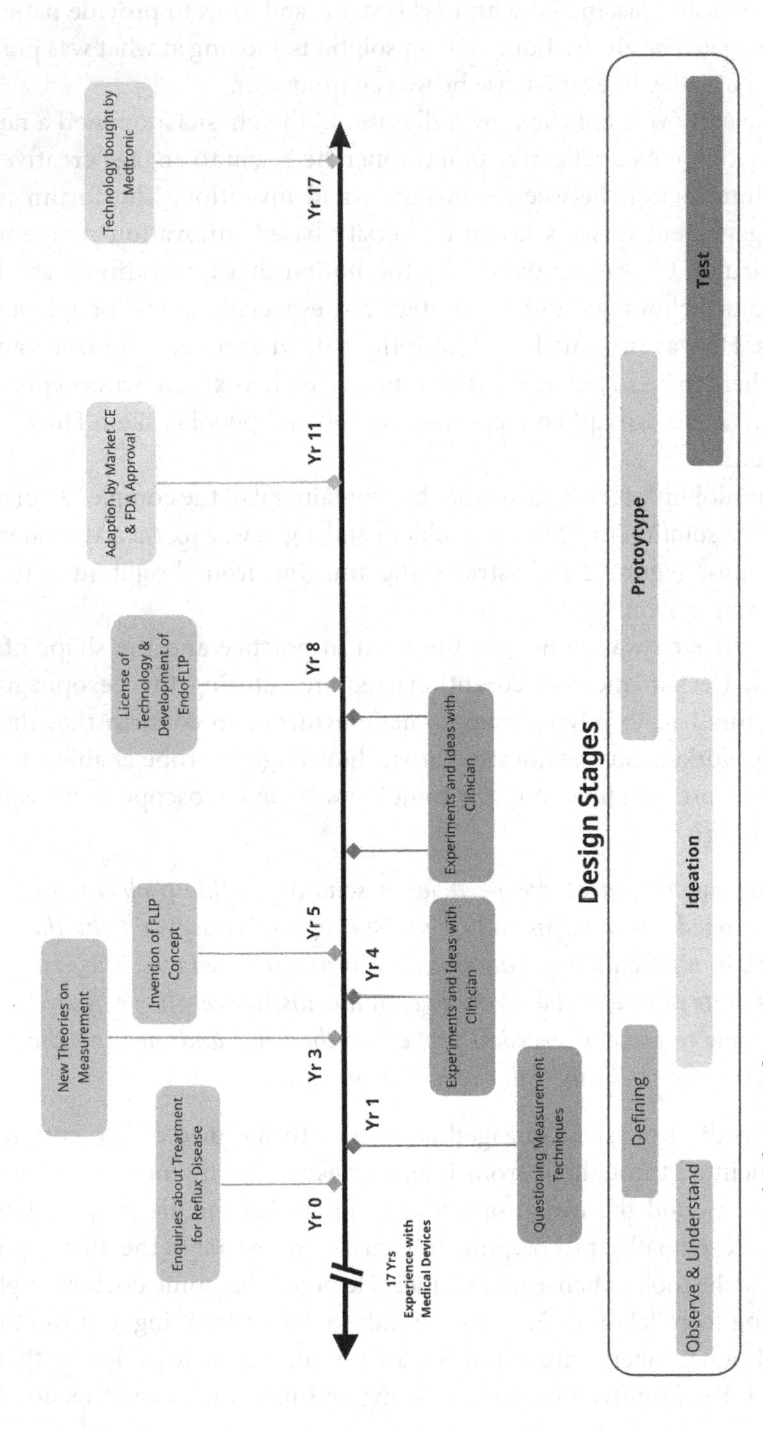

FIGURE 2.2 EndoFlip® timeline.

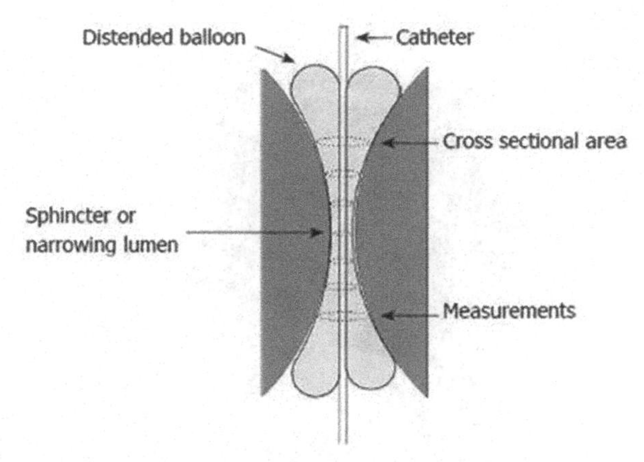

FIGURE 2.3 Placing the probe.

He found it to be illuminating. On reflection, he could see connections between his behaviour and actions with fear, bravery, courage and risk.

> *Based on our interaction with users and the idea that we wanted a practical probe that could be used to give real-time data, we didn't settle with that but went a further step to create a probe to give a 3-dimensional representation of the area of the valve region. This image we referred to as the "functional lumen imaging probe" or FLIP. Subsequently, we commercialised the medical device as EndoFLIP®.*

Such an innovation would require certification and the early steps were taken with ethics approval in Denmark and in other small studies. Some of these studies were on animal models. The studies built a platform from which the innovator could consider eventual realisation of a product which could be awarded a CE in Europe and FDA approval in the US – both essential for any subsequent sales.

As the work progressed, the medical physicist decided to consider exploiting the idea in a company in Ireland. Over two days, he met with seven venture capitalists. All expressed interest but nobody funded the idea to its next step. Then, there was a breakthrough. While meeting with his Danish co-inventor, they received an invitation to visit with a medical company in Ireland. During the visit, the company asked about reflux. A chance connection with a highly astute, enabling individual led to an introduction to entrepreneurs working in the medical device start-up scene in Ireland. The product emerged and Crospon, a start-up, was formed.

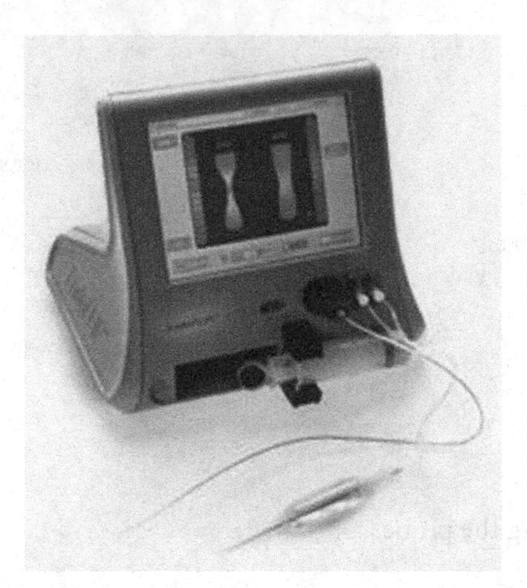

FIGURE 2.4 EndoFlip®.

The first Endoflip® device was later bought from the start-up by Medtronic (Figure 2.4). EndoFlip® is now a family of products. Originally used for stitching, new applications are emerging in surgical cutting and other measures of diameter and volume in the gut. The probes are linked by a common trait – the ability to assess taut circular muscles of the sort you often find in the digestive tract. Medical practitioners can use a probe to measure how much to cut muscle and other tissue until the best therapeutic effect is achieved. They have much better geometric information on shape and structure in hollow organs, particularly those with muscle function. The family is at the centre of a multimillion-dollar-valued offering to the medical device market and is used worldwide by doctors and surgeons to assist with the diagnosis and treatment of a number of key digestive disorders.

2.4 CONCLUSION: SO WHAT? WHY NOT?

What do these cases show? First, creative thinking, design and innovation are possible. Second, they demonstrate a mindset which is open to new ways of thinking, reflecting and learning. Third, they benefit from accommodation and integration of different perspectives and voices. Without such engagement, progress is limited.

These case examples set the scene for our focus on the healthcare sector in this book. Creative thinking, design and innovation are evident already in the sector. However, there is evidence also of the "Red Queen Effect".

Here, target levels of improvement objectives will not necessarily remain constant. They can and do shift in dynamic environments. As in Lewis Carroll's *Through the Looking-Glass*, "…it takes all the running you can do to keep in the same place. If you want to get somewhere else, you must run at least twice as fast as that!" Societal expectations, availability of new treatments, resource availability and political decisions combine to create a dynamic within which the improvements of today cannot or do not deliver. In addition, shocks like Covid-19 disrupt and divert from the more routine processes behind the provision of healthcare. So, the application of creative thinking, design and innovation may have to deal with multiple and unexpected challenges. That does not mean that such activity should stop or simply address the demands of the current crisis. Instead, they should be undertaken in a way that innovates for today and tomorrow. Leveraging change in this way to solve complex problems and challenges can be more efficient and effective as well as inclusive and engaging for those who are involved.

SOURCES AND SUGGESTED FURTHER READING

Coughlan, P. & Coghlan, D. 2016. Action Research in, Christer Karlsson (editor), *Research Methods for Operations Management*, Abingdon OXON, Routledge, 233–267.

Dell'Era, C., Magistretti, S., Cautela, C., Verganti, R. & Zurlo, F. 2020. Four kinds of design thinking: From ideating to making, engaging, and criticizing. *Creativity and Innovation Management*, 29, 324–344. https://doi.org/10.1111/caim.12353

IDEO Shopping Cart Challenge: https://www.youtube.com/watch?v=M66ZU2PCIcM

Jobst, B., Köppen, E., Lindberg, T., Moritz, J., Rhinow, H. & Meinel, C. 2012. The Faith-Factor in Design Thinking: Creative Confidence through Education at the Design Thinking Schools Potsdam and Stanford? in, H. Plattner et al. (editors), *Design Thinking Research, Understanding Innovation*, Berlin Heidelberg, Springer-Verlag. https://doi.org/10.1007/978-3-642-31991-4_3

Kelley, T. & Kelley, D. 2012. Reclaim your creative confidence. *Harvard Business Review*, 90(12), 115–118.

CHAPTER 3

Developing the Potential of Healthcare Professionals

IN HEALTHCARE WE KNOW that the delivery of care is not just about doctors and nurses. For professionals and readers from outside healthcare, an example of some of the roles needed to deliver healthcare is shown in Table 3.1.

This list doesn't include sub-specialties, particularly in medicine and nursing, of which there are many.

Often healthcare workers can have complex working relationships. They may report to a more senior person in their profession but, if they are working in a multidisciplinary team, they will often have a separate reporting line to the team leader or head of the group. This duality can be difficult and cause confusion and divided loyalties at times. However, as we have seen the world over, especially during 2020 through 2022 with the Covid-19 pandemic, often, these workers are at the core of delivering healthcare in society.

Together, healthcare professionals must perform day-in day-out facing whatever medical dilemmas are thrown at them. Not only are they responsible for the knowledge they must possess and the treatment they provide but they are responsible for the quality of care, speed of delivery, dependability of service, flexibility in their working lives and also delivering this level of service within an ever more constraint cost environment. Of

DOI: 10.1201/9781003123910-3

TABLE 3.1 Roles to Deliver Healthcare

Doctors (in many disciplines)	Clinical Engineers	Dieticians
Nurses (in many disciplines)	Medical Physicists	Audiologists
Pharmacists	Radiographers	Laboratory Scientists
Speech Therapists	Occupational Therapists	Bioengineers
Physiotherapists	Physiologists	Healthcare Managers

course, even more than having accountability to their colleagues, management and their profession, they must also be accountable to the patients they individually and collectively care for. In this complex or multifaceted, highly regulated and sophisticated environment, they must support each other, avoid mistakes and comply with policies and procedures. It can make for a difficult environment day by day, and it can be extremely hard for these workers to step back from the delivery, reflect on what they are doing, be creative and innovate.

Yet, we contend that it is this very practical process of innovation that will dig them out of the system and process problems they encounter and is the best option to be more efficient and effective in their workplace.

3.1 DEALING WITH MULTIFACETED PROBLEMS AND CHALLENGES FROM THE HEALTHCARE PROFESSION PERSPECTIVE

You could say that, for the most part, delivering healthcare is a problem-solving exercise. As we can see, it is not just the doctor acting alone who is involved in the diagnosis, treatment (or treatments), followed up by care and monitoring until the patient is 100% well or until their chronic condition is fully under control. However, many of the contributing professions can see problems or challenges with providing this care, especially if it requires action outside of their core professional roles. It's "not their jobs" or "a distraction from the important work they need to do".

So, even if the job of providing the tools needed to deliver healthcare such as treatment rooms, clinic booking systems, waiting lists or medical device set ups are someone else's responsibility, we need to consider what happens when things go wrong. Examples may be:

- A certain disorder type becomes more prevalent very quickly and the clinic used to provide for the patients becomes over-run with patients.

- In a procedure room which is very dependent on technology, such as a magnetic resonance imaging scanner, the equipment fails suddenly, is not repairable and will not be replaced for six months. How will the patients receive their care?

Increasingly as medicine and clinical care become more technologically intricate, the problem-solving techniques needed to troubleshoot them will also become more complex. Healthcare professionals cannot assume that the management team, the administration department or the responsible government healthcare division has the knowledge and skills to solve the problem. As technology is used increasingly to solve systemic or data management problems, we need to consider how we can ideate and use it to best advantage in healthcare. So, we often make the mistake of assuming;

- The managers will fix the problem.
- The IT team will design an app or software work around to fix it.
- We can hire consultants to fix this problem.
- There is an off the shelf computer solution to help.
- Everything will be sorted when we move into the new building.

However, this type of approach is not needs based. The practitioners and professionals working in the area where the particular problem lies have the knowledge, information and experience to be fully aware of the problem that needs to be addressed. This awareness must translate into developing a meaningful and articulated statement of the problem.

Pushing a solution to a systematic problem in healthcare on a professional group will never be more than partly successful. However, using some workshopping techniques with a mix of the experienced professionals involved and other smart and enthusiastic people provides a better way to find a solution to the problem or challenge based on need.

Realistically this approach can only happen when:

- We provide an environment and atmosphere where people feel it is safe to be creative and to share and develop ideas.
- Accepting that no one person is likely to have the answer to the problem.

- Sharing ideas and working together using techniques related to gamification and playfulness to help a team get closer to the best solution to the problem.
- The healthcare organisation is set up to support staff to run workshops, give permission to execute the process and then review and allow the ideas that come forward to be used to address the problems that are being confronted.
- Also, for the process to really work, a range of staff at all levels and grades forms part of the workshopping team and the engagement process is structured so that all have equal input.

3.1.1 Re-balancing the Status Quo When Things Go Wrong

Providing an environment where a diverse group of professional staff can develop ideas to solve multifaceted problems is often not innate in healthcare. Further, building models or prototypes of what the solution might look like is not intrinsic. Trying-out new group-developed ideas and concepts towards finding a better solution is a relatively new way to work in healthcare. In fact, changing the way we deliver medicine or the wider tenet of healthcare requires a process to be in place so as not to introduce the new plan or idea unless it is fully tested. This rigor is required even if, for example, we were trying an initiative not directly related to patient care, such as developing a new way to organise a patient clinic. If the new method fails, it could have severe negative implications for patients such as delayed diagnosis or treatment or even worse. This sensitivity can create a fear factor for healthcare workers who might otherwise want to get involved. It can have a negative impact on their thoughts and on the creative thinking about how to change or improve the process. Allied to this sensitivity is that healthcare professionals can often see process problems as someone else's responsibility, such as administration or management. Relatedly, they can perceive the work they are doing in providing clinical care and related activities as separate and, even, more important. In the view of the authors, this is an incorrect perception.

3.2 WHAT IS IT ABOUT HEALTHCARE PROFESSIONS?

Our socially responsible yearning to provide accurate diagnosis, ideal treatment, optimal care and restored health has led us to believe that developing highly educated and trained healthcare professionals such as doctors, nurses, pharmacists and physiotherapists, to name a few, is the best way forward. Broadly speaking, this development and professionalisation of healthcare has allowed us to live a full 26 years longer than we did 100 years ago (United Nations). However, as healthcare becomes more sophisticated, it also becomes more complex. The digital world and the overall rise in the level and standard of education have created a need and yearning for better healthcare services throughout the developed countries and beyond.

So, where once the healthcare centre or hospital was seen as a place for the doctor to diagnose you, treat you and the nurse cared for you, now these institutions are sophisticated temples involving intricate and sophisticated technologies and as many as up to 100 speciality and subspeciality professions. The result can be confusing, especially for the public (*Irish Times*, November 22, 2022).

This can be all fine when things are going well but what happens when there is a problem in the system? This could be as simple as the technology that is used in a medical process breaks or as complex as an entirely new treatment becoming popular and the way the service is delivered to patients needs to change. In our world, or in systems, in general, these are referred to as multifaceted problems.

However, all the training that healthcare professionals get and all the education they complete does not often prepare them for the problems and challenges they face in delivering the best care to patients. These problems are most often process problems. Knowledge and experience within a specialist area of medicine do not always prepare the professional for these problems and challenges. Also, the problems and challenges change on a regular basis so having one formula or pathway to fix a problem or challenge might not be useful for the next problem or challenge.

3.3 STRUCTURE OF HEALTHCARE ORGANISATIONS AND WHY THEY DON'T FOSTER INNOVATION

3.3.1 Pecking Order in the Healthcare Professions

There cannot be a conversation about interdisciplinary working in healthcare without touching on the issue of hierarchy. In many, if not most, developed countries and beyond there is still a professional pecking order in healthcare. With regard to medical decisions, the rule of thumb is typically that the senior or consultant doctor's or senior nurse's opinion takes precedence, and they make the ultimate decision as to which diagnosis, treatment and care the patient gets. So often when there is a discussion about multifaceted problems related to healthcare delivery, this precedence exists or is the default position. Yet, this established approach is inappropriate when solving this type of problem, especially where the most senior team member may not have the best ideas on how to solve a multifaceted problem.

The established approach is evident when there is a series of problem-solving meetings, often led by the most senior doctor involved with the team. The hierarchical structure facilitates behaviours such as:

- Senior staff over articulating their position and unduly influencing the proposed solution to align with their perceived priorities.
- Senior staff not turning up for many or any of the formative meetings, waiting until the end of the process and then, vetoing the emerging decisions as they do not align with their perceived priorities.
- Senior staff not getting involved on any level in the problem-solving process because they are too busy, and then complaining when the solution is not working.

3.3.2 Silo Mentality

While specialist healthcare professionals are highly educated and highly trained, much of their work culture, mindset and identity can be related to their profession. Further, their opportunities for promotion and advancement are related to how they represent themselves within the bounds of their profession. So, while, on the face of it, many professionals working in medicine do so in a collaborative way, there may not be motivating factors or incentives in the workplace to support collaboration. For example, what happens when a diverse medical team makes a mistake where there is a patient incident, or an error causes a patient's death? Individuals, fearing blame, may return to their professional group seeking support. So, at the very time when a severe problem or challenge needs to be addressed, instead of staying together to investigate and resolve it, the work culture is such that individuals run back to the safety of their own profession. Taking a collaborative problem-solving approach to ideating towards a workable solution and implementing a plan to execute that solution is often better and faster. It contrasts with a silo mentality that precludes a collaborative, problem-solving approach. Such a mentality is not new in organisations or businesses, but it is evident in healthcare where professionals are driven by their particular professional standards and allegiances and, in some cases, the fear of litigation. Such behaviour can often make highly skilled professionals slower to adapt to new ways of working and resolving problems outside of the strenuous demands of their core training and professions. If you are interested in further reading on the topic of silo thinking, consider reading Gillian Tett's excellent book *The Silo Effect* (2016).

3.4 STRUCTURE OF HEALTHCARE ORGANISATIONS AND THEIR ABILITY TO FOSTER INNOVATION

3.4.1 Multidisciplinary Team versus Interdisciplinary Collaboration

Where healthcare professions focus on the problem from their professional standpoint rather than a collaborative interdisciplinary group perspective, a silo mentality can be the reason why it is difficult to solve in an efficient and innovative way. Yet, in a world where concepts of care are directly related to medical technology and computer and information technology as much as medical expertise and knowledge, this silo mentality creates a barrier. Multidisciplinary teams are often unable to develop a cohesive approach to a multifaceted challenge, as each team member uses their expertise to give a narrow approach to the problem. In contrast, interdisciplinary teams build on each other's expertise, their mutual respect and trust to develop common shared goals and outcomes. This interdisciplinary approach is usually not innate in healthcare. However, with a small amount of training, that we usually refer to as upskilling, and the right facilitation, healthcare professionals can easily learn it.

3.4.2 Changing the Approach to Engagement, Coordination and Decision-Making: Meetings versus Workshops

Most healthcare workers start their working lives without training in facing problems and challenges outside of their professional training when delivering care. So, if there are issues with healthcare delivery which they perceive as being outside of their professional training, or they just want to discuss ongoing work challenges and activities with the wider team, even in a very loosely managed environment, their default activity is to arrange and attend a meeting.

Meetings can serve a purpose and can be good for:

- Instructing staff.
- Communicating information from management to staff and from staff to management.

However, meetings are not good for:

- Creating a shared understanding of the problem at hand. Often problems are misunderstood, misrepresented or, even worse, interpreted differently by different people in the room at the same time.

In contrast, using well-facilitated and proven workshop techniques, particularly those related to innovation staff, can:

- Become better able to understand problems by defining them in terms of the user and the user's experience.
- Create spaces and activities where those in the room can generate ideas together, explore those ideas and filter through them.
- Draw on the collective experience of people in the room to produce the best or a better solution to the problem or challenge that is faced.

Very often in healthcare, professionals meet and discuss issues without any real facilitation skills at play. This shortfall often results in only the louder and more articulate voices being heard, actions and decisions not recorded and, if they are, individuals not allocated responsibility for actioning items. And then, actions are not completed because everyone is too busy. Unproductive, even unstructured, meetings can run for too long and achieve very little. Attendees at these meetings can quickly get frustrated and either fail to turn up at further meetings or are not motivated to contribute. Ultimately, there can be no respect for or trust in the meeting process where needed contributions are not made, and potential is not harnessed. In a nutshell, some individuals with contributions to make become disengaged and remain silent, believing that their voice is not important and they are not valued.

3.4.3 Diversity in Interdisciplinary Teams

It is not instinctive in healthcare to involve other diverse disciplines in problem-solving exercises. In fact, we often think we should have only the experts in the room. These experts are usually several senior doctors and nurses and, perhaps, other senior healthcare professionals. Yet, research and experience would show that to produce a creative solution to a multi-faceted problem, you need a broad range of people from diverse education and experience backgrounds in the room. A diverse group including people who are smart and creative but who are not necessarily experts in the problem can help the team to generate novel ideas and, so, challenge the wider group thinking. Even if the problem or challenge is seen as primarily a nursing one or a medical one, it is possible that evidence of a multifaceted problem will emerge and, so, will benefit from a diverse group of engaged people trying to solve it.

Said differently, having a homogeneous group of people, such as all nurses or all doctors, who think alike trying to solve the problem or challenge is not advisable. A group of people who tend to think alike are more likely to set limitations on the ideas they might think of and therefore be less creative. Having some people in the room who are smart, bright and engaged but who are not subject experts can provoke and engage new and disruptive ideas towards solving problems.

3.4.4 Permission from Management

Behaving collaboratively and using innovation practices to try to develop and advance better solutions to problems in delivering healthcare can challenge a healthcare organisation. It is important that senior executives and management understand what is going on and support it. If healthcare professionals and other allied staff workshop new ways of working and solving problems using innovation practices, then it is critical that they have permission from management to act on these new solutions. If some of their worktime is being spent attending workshops and, in some cases, these workshops may use playful and gamified techniques, it is important that the management team recognise this approach, approve it and support it.

As facilitators, when we work in this way, we consider it necessary to bring this need for permission to the attention of the management early on, usually through engagement with the executive team. Even better, we also find it relevant to let the senior team experience it. If you describe to senior management that you are going to take 24 healthcare workers out of their regular jobs for three or four hours and you are going to teach them how to solve problems using games and play techniques, they might think you are crazy. However, if you set up a workshop for executive members, bring them on a facilitated journey through some of the techniques that are to be used, they usually get it, and what is in prospect for their colleagues will make a lot more sense to them. Even if, at this point, they may not completely buy into the process, it will usually lead to permission to set up and run workshops which can demonstrate the usefulness of innovation practice in solving real-world problems.

3.5 HOW DO WE GO ABOUT CHANGING THIS APPROACH?

3.5.1 Needs-Based Approach

Through one of the author's lengthy career as an engineering scientist working in healthcare, he noticed the number of products and technologies that were launched into the healthcare market that may have been amazing in terms of their uniqueness but were not developed in a way that made them optimal for use in practice. While this situation has improved in the last two decades, it still has a way to go. One simple example is the number of medical devices with options and functions on them that are never used by healthcare staff. What you often find is that devices are developed by engineers to solve a functional problem in the healthcare space such as how a particular parameter will be measured. These engineers can be one step removed from delivery of healthcare and, therefore, develop devices they think healthcare professionals need, rather than asking the professionals about their needs or the needs of their patients. In our experience, few device developers take the time to understand the clinical practice that is going on, how the healthcare professionals are looking at this practice in terms of a process to deliver care, and then developing or demonstrating an innovative technology in terms of the user needs. Invariably this can often mean that a new technology is never the best fit for the need and the healthcare staff, as users, must perform workarounds to get the device to work optimally to suit their precise need. Understanding the need and then developing the design and science of the technology with a specific user focus can be key and help develop the best products for patients and healthcare.

3.5.2 Learning from Other Industries

It is important that those working in the healthcare sector do not miss the opportunity to learn from other industries. In our case, and in the case of this book, it is the use of and experimentation with innovation tools and techniques used by other industry sectors, such as the digital industries, that have informed the adaptation and use of the techniques laid out in this book.

While many of the techniques focus on new product development and process improvement, we believe that with research, small adjustments and the application of common sense, they can be adapted and focussed on the problems and challenges in delivering change and creating improvements in healthcare. In a sense, this is what this book and the skill sessions and workshops we run are all about. So, it is not just theory we apply, it is applying it in the real-world setting, to solve real problems for real patients and healthcare professionals in the healthcare setting.

3.5.3 Building Trust

Healthcare professionals can be wary of management. Often there is a "them and us" attitude. They can see attempts by the powers-that-be to improve efficiencies and activities as distractions from their core professional work. Often, they will even go as far as to see healthcare delivery as someone else's responsibility and not their problem. The idea that they can just arrive and expect to provide the service they were trained for appeals to them. They may not see it as part of their responsibility to manage, for example, the "handling" of patient flows. Rather, they see it as someone else's job to set up and manage the multifaceted processes and systems that bring patients to them for the service they provide, be it care, diagnosis, treatment or monitoring.

Also, sometimes, when they do commit to a "not that well thought out" change process and they have a terrible experience, they become distrustful of management to the point where they will be reluctant to get involved in or try out a new method again. Building trust takes time and patience. It requires skilful facilitation and a certain amount of leadership from those who will clear the way for innovation and bring practice to the fore.

3.5.4 Demonstrating Behaviour

We now know that we can bring methods into play in the healthcare delivery arena that are more creative, more disruptive and provide more efficient and effective solutions to ongoing problems. Very often the best way to get sceptical professional staff to engage is to demonstrate how it is done. In many respects, as innovation practice is a skill, it is best demonstrated by action. As facilitators, when we first work with a new healthcare team, we try to work with the executive team in the initial stages. In fact, we get them to experience a workshop in a problem or challenge they face. Once they have some experience with it, they will usually support its roll-out to other staff. If you can say to healthcare professionals that these are techniques that we have introduced to management, they liked them, they found them useful and they supported them, this gives a good basis for people in healthcare to be a little less sceptical and more open to having a go.

Once engaged, the early workshops with those new to the methods tend towards working on problems and challenges that are more open and, in a sense, easier to ideate and brainstorm around. Early workshops with these groups are structured to both (1) work on problems that are live and real to the participants and (2) have an element of training built in so that staff attending can learn from the experience and start to use and experiment with concepts they have seen when they go back to working in their own departments and specialities.

3.5.5 Convincing Healthcare Workers to Turn Up and Invest Time

Getting workers to turn up and invest time in the workshops can be difficult. Again, the link with the management staff can be helpful. As facilitators, we often set up a structure where many of the higher-level problems and challenges we work on have a sponsor from the management team. The sponsor may or may not directly take part in the workshopping sessions, but are valuable in that they:

1. Demonstrate permissions from superiors to ensure staff are released from duties to attend workshops.
2. Provide information and data on why the particular problem or challenge is important for the organisation and assure them that the commitment for the work on it is supported by senior management.
3. Provide other valuable information (such as on constraints) that may not be available to the workshopping team. For example, in some of the work we talk about in this book, we were planning services for a new children's hospital. However, although we were trying to be open, creative and disruptive in planning how services would run in this new entity, we did have to work with the constraint that the building was already designed and under construction. So, we needed to design processes and work practices that would fit in with this design.

3.5.6 Building Champions

In healthcare, developing service in a new speciality or treatment usually starts with the key individuals getting specialist training, then setting up the service and developing it slowly as the skills and experience develop among the medical and healthcare teams. In contrast, rolling out innovation practice to solve healthcare problems does not follow this path. Taking

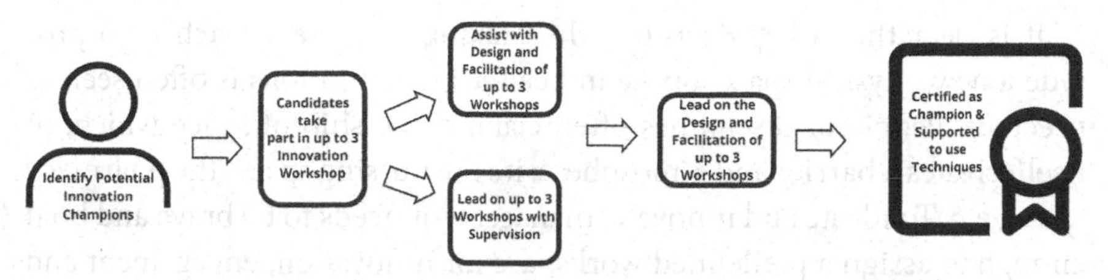

FIGURE 3.1 Key skills to become an innovation practice champion.

an action learning approach, in essence, learning by doing, questioning and reflecting is more appropriate. With the right facilitator, the participants can learn in action, questioning and reflecting on the outcomes.

Further, we have found that healthcare professionals who are more receptive to such techniques will engage with them faster. So, the real skill for the facilitators is to identify these individuals, then encourage and guide them in using the tools while developing their own skills in facilitation.

Figure 3.1 identifies some of the key skills that healthcare staff can acquire to become innovation practice champions in a healthcare organisation.

3.6 CONCLUSION

It is embedded in healthcare work that if you want to do something differently or require new knowledge and skills, you must train formally in it first. Most healthcare workers are heavily laden with formal education, many trained to postgraduate level. We believe this is not the necessary approach for the roll-out of innovation practice. Instead, upskilling rather than formal training is necessary. We believe that with a modest, interactive intervention they can add some practical innovation tools and skills to their armoury and that this can assist them in problem-solving in the healthcare environment.

Thus, learning-by-doing forms the basis of our workshopping and allows us to take on real challenges and problems in a learning environment where teams work on real problems faced in their healthcare workplace.

It is important to find such ways in healthcare to break down these professional barriers which, though important for professional standards, are counterproductive for interdisciplinary problem-solving. Healthcare organisations need to invest time and effort in generating new pathways in the healthcare workplace where staff from all areas of healthcare can come together and generate new ideas for solving challenging workplace problems.

It is clear that a key decision when taking a new approach is to provide a new physical space. Space in healthcare institutions is often seen as precious. People or disciplines often claim ownership of space which, in itself, creates a barrier and atmosphere: it's the nursing space, the pharmacy space, etc. To ideate and innovate, management needs to be brave and bold enough to assign a predefined workspace for innovation, engagement and collaboration.

SOURCES AND SUGGESTED FURTHER READING

Department of Economic and Social Affairs, United Nations. 2021. United Nations Reports "Global Population Growth and Sustainable Development." Available at www.un.org.development.desa.pd/files/undesa_pd_2022_global_population_growth.pdf

Tett, G. 2016. *The silo effect: The peril of expertise and the promise of breaking down barriers*, New York, Simon & Schuster.

Thompson, S. November 22, 2022. Why do we try to confuse people who come to an Irish hospital? *Irish Times*.

CHAPTER 4

Understanding Multifaceted Problems and Challenges in Healthcare and Setting about Solving Them

4.1 DEFINING CHALLENGES AS PROBLEMS

Many challenges are misdefined or pre-defined as "problems". Here, management or senior staff might throw the challenge to their team or teams and ask them to set about finding a solution. The team then engages in the assigned task but without conscious consideration of the nature of the challenge or use of appropriate tools. Yet, it is recognised that a clearer definition gives everyone a shared understanding of the challenge and leads to a better solution. Further, there is no question that collaborative teams using innovation tools can help healthcare professionals to respond better to challenges through developing a shared understanding and being clearer about how they might go about finding a solution. However, this summary of good practice leaves undefined the concept of a problem. Said differently, what is a problem? If we know what it is (and is not), then we can go about solving it.

In normal discourse, many challenges are labelled as problems. As such, problem and challenge may be seen as synonymous. However, such

DOI: 10.1201/9781003123910-4

colloquial use of language is not helpful, especially when clarity of definition can be empowering. Let's look at the following statement:

"Waits and delays are not only disrespectful to patients, they are also potentially harmful. The right patient care must occur at the right time and in the right place."

So, what is the challenge here? Is it an absence of respect for patients and their welfare? Is it an absence of understanding of the link between waits and welfare? Is it a shortcoming in patient tracking to enable tracking where the patient is, what care they need and when? We could go on. The point is, however, that the short statement prompts consideration of categorically different questions, each attempting to diagnose the cause so that a solution can be found. Consider each of the three diagnoses outlined:

- If it is an absence of respect for patients and their welfare, what is the challenge? Is it an absence of basic humanity, an absence of training or an absence of empathy? A response to each of these emergent questions might take the challenge in a different direction: pick the right people, or train them appropriately, or encourage staff to take time to connect with the patient and appreciate the situation from the patient's perspective. Or, maybe, it is a combination of all three. Whichever pathway is taken, the solution might not be guaranteed. So, why might that be?
- If it is an absence of understanding of the link between waits and welfare, what is the challenge? Is it an absence of understanding between action (waiting) and outcome (welfare)? A response to this question might take the challenge into defining appropriate metrics for waiting times and welfare, exploring the measurement system which gathers relevant data, the timing of data availability to those responsible for the process or interpreting the emergent data. Or, again, it may be a combination of all three.
- If it is a shortcoming in patient tracking, what is the challenge? Is it the absence of a functioning system to identify and locate patients in a timely fashion? Is it an unreliable data management system to enable knowing where the patient is, what care they need and when? Or is it an uncontrolled and uncoordinated response to an unreliable system where individual staff try to compensate for poor tracking data but cannot keep up? Yet again, it may be a combination of all three.

Here then is where our definition of "problem" becomes relevant. As with all such definitions, it is helpful to define a problem in comparison to a challenge which is not, categorically, a problem. Said differently, not every challenge may be a problem – it may be a puzzle. The challenge may contribute to disruption or poor performance. It may lead to resource wastage or poor patient outcomes. However, if we err in categorising the challenge, we may waste time, effort and energy looking for a solution which is either unachievable or simply found.

To understand the distinction between problem and puzzle, we turn to action learning. The philosophy of action learning runs through this book. In Chapter 7, we deal with it specifically. In brief, action learning (Coughlan & Coghlan, 2011) is an approach to learning in action where individuals with a commitment to action and learning form a group to address a problem for which there is no single or technical solution. The link to this chapter is in the final part of that sentence: *a problem for which there is no single or technical solution*. The inference is that certain challenges have a single or technical solution, while others do not. Further, the relevance of action learning is limited to those challenges for which there is no single or technical solution. So, at the heart of action learning is a distinction between and among different kinds of challenge.

Reg Revans, the "father" of action learning, distinguished between puzzles and problems (Revans, 1998). Puzzles are those challenges which are amenable to specialist advice and for which a correct or single solution exists. Referring back to the earlier statement about waits, delays and patient welfare, an example of a puzzle would be a technical failure of equipment or a tracking system. In response, a specialist technician may diagnose the failure mode, replace the defective component, re-programme the system and re-start the equipment. Puzzle solved without creativity or innovation!

Problems, on the other hand, are challenges where a single solution cannot possibly exist. Most complex organisational challenges fall into the category of a problem, as there is no single solution but likely to be many opinions as to what a preferred course of action might be. Different people, within or outside of the function or discipline, can advocate alternative courses of action informed by their individual value systems, past experience and perception of the desired outcomes. Referring back to the earlier statement about waits, delays and patient welfare, an underlying problem might be the uncontrolled and uncoordinated response to an unreliable system where individual staff try to compensate for poor tracking data but cannot keep up. In response, the staff, who are professionally committed to taking action, may need to develop confidence to question others in the system while sharing their creative ideas so as to learn about (and from) the problem. And, even if committed and ideas are shared, they may not get it right – which may uncover a new problem. Their colleagues might not be literate in the language of their diagnosis, open to hearing an alternative diagnosis or collaborative in responding to the diagnosis.

4.2 FINDING SOLUTIONS

With this understanding of the characteristics of a problem, it follows that solutions can be disruptive. In contrast, the solution to a puzzle may result in a return to the status quo. Of course, if the puzzle keeps occurring then, as in the immortal words of the crew of Apollo 13, it becomes clear that "*Houston, we have a problem!*"

At this point, tools and techniques such as design thinking, game storming and creative thinking can come to the fore to help healthcare professionals to imagine and find novel solutions. Details of these approaches are explored in other chapters. For now, the search for solutions may take staff outside of their normal scope of activity and into contact with colleagues from other functions or specialisms, or patients and their family members. These problem-solving actions might require taking on new levels of responsibility or sharing responsibility with others in a novel way. So, finding disruptive solutions may require getting teams into an environment where that can ideate and be creative in their idea generation, experimentation and implementation. These actions might require the healthcare professionals to learn and upskill.

Similarly, solutions may require incremental or radical change. With incremental change, the assumption is that small and progressive improvement steps can simplify a process and improve performance. Extending incremental change from a single initiative into a series of connected actions can form the basis for continuous improvement, harvesting "low-hanging fruit". Continuous improvement may yield a series of incremental improvements and, in contrast to radical changes, they may be less disruptive and lead to other follow-on improvements. As such, continuous improvement can emerge as a "natural" way of questioning, collaborating, experimenting, taking action and evaluating within the operation.

In contrast, disruptive solutions can emerge when the problem-solving leads to major and dramatic change in the way the healthcare operation works. The solution may represent a step change in practice and lead to outcomes which are sudden, often irreversible but, hopefully, an improvement in performance. Disruptive solutions are rarely inexpensive and may require investment of time and resources. They may disrupt routine operations and require process, staff, technology and procedural changes. Further, as an approach to improvement, the benefits of disruptive solutions may be difficult to realise quickly. Ultimately, disruptive changes can lead to more radical improvements in efficiency, effectiveness and resource saving.

4.3 THINKING AND ACTING STRATEGICALLY

There is a danger that firefighting puzzles and problems (and even collaborative problem-solving) can come to dominate the time and space for action. In contrast, good practice would suggest standing back beforehand, thinking and acting strategically to set the direction and to frame the planned improvement initiatives. Linking them to the goals of the broader organisation will lead to a clear distinction between discrete projects and a programme of creativity and innovation. As individuals, healthcare professionals work within a team or department, and that department may exist within a hospital facility which is part of a group contributing to regional or national delivery of healthcare objectives. As such, initiatives developed to address local problems may have the potential to demonstrate actions to others within the same facility, region or nationally. Said differently, addressing local problems may have the potential to maintain the local status quo and to achieve parity of performance with other teams within the same facility. However, they may also provide credible support towards achieving established regional or national goals, or demonstrate a novel contribution and new thinking towards setting new regional and national goals.

A practical example of such thinking and acting comes from our interaction with a hospital pharmacist. The pharmacist identified a range of shortcomings (and, so, opportunities for improvement) in the dispensing process. These shortcomings included delays in dispensing urgent medication requests, contacting the correct person to resolve a query on a medication item, waiting on a “special order form” where the medication had initially been requested on the wrong form, responding to cases where medication items were short, delayed or unavailable. In response, the pharmacist engaged with her team of pharmacists and technicians, firstly to outline and analyse critically the current process for ordering and dispensing medication. Arising from the analysis, the pharmacist detailed a set of specific and actionable recommendations to improve the process with associated key performance indicators. The expected outcome was a move to a more streamlined process of ordering medication, with less wasted time, fewer forms and the introduction of a local safety check channelling emergent queries through the ward pharmacist. Implementing the recommendations on two wards was planned as a pilot to gauge feasibility ahead of a hoped-for hospital-wide roll-out. However, when the pharmacist presented the initiative to a mixed group of healthcare professionals from outside of her hospital, the initiative resonated with their experience as pharmacists, nurses and clinicians. There was clear recognition of the problem identified and the wider potential of this local initiative to effect improvements beyond the original hospital pharmacy setting. Effectively, the strategic potential of the initiative emerged, and a new improvement cycle began.

4.4 GETTING THERE THROUGH PROCESS INNOVATION

Over recent years, we have worked with a number of groups of healthcare practitioners, all engaged in developing an understanding of creative thinking and innovation through participation in a graduate-level programme. One module focuses on process innovation. As a short pre-module assignment, each participant is asked to describe briefly a process innovation opportunity in their own area of managerial responsibility. They should:

1. Describe the process and identify the contributors to process performance.
2. Identify the choices in managing, structuring and running the process, and integrating different functional capabilities during the process.
3. Explore potential actions to improve the process.

At the end of the module, participants are invited to revisit and develop further the process innovation opportunity identified in the pre-module proposal. At this point, they should:

1. Revisit their description of the process and identify the contributors to process performance.

2. Revisit their identification of the choices in managing, structuring and running the process, and integrating differing functional capabilities during the process.
3. Explore specific and actionable recommendations to improve the process including a description of the proposed action, time scale for introduction, resources required, knock-on effects anticipated in other areas, measures of performance of the change and expected outcomes.

The following prompts facilitate their understanding of the multifaceted problems and challenges associated with the focal process and setting about solving them:

Prompt 1: *What moves through the process from start to finish?*

Guiding questions:

How does the delivered service reach its finished form? What is converted into what? What are the elements of work involved in getting from start to finish, in creating the service which goes out the door? What flows into and out of the area per day? How many different types of people, materials or information flow in and out?

The approach requires the healthcare professionals to include in their observations the flows of people, materials or information; the apparent checks carried out; the contributions of equipment and information systems at each stage; the apparent attention paid to timing and sequencing; and the way people, materials and information are joined together or separated out.

For this first prompt, many find it useful to draw a process flow chart to illustrate what happens and locate problems involved in running the process. The guiding questions invite the professionals to consider what moves and to envisage how many different things have to be managed and handled every day.

Prompt 2: *How is the process organised to be effective in its performance?*

Guiding questions:

- What aspects of the working process make the entire process effective – the size of the equipment or getting a long run on similar activities and thus saving on set-ups?

- What is the function of any stopping points where work-in-process accumulates? Is it to accommodate anticipated variations in flows or because the next stage in the process is not reliably available? Here, think about how the resources of people, space, materials and equipment are organised to function in practice.

Prompts 1 and 2 require careful observation and analysis of the physical organisation of the system. Considering each question may lead to yet more questions. A separate line of questioning, which is worth pursuing, involves matters of judgement rather than matters of observation. What drives a process is not what can be seen on the floor in terms of layout, equipment and piles of work-in-process. Rather, it may be fruitful to consider how the procedures, targets and pressures, together with the motivation of people in all the related locations and functions, combine to drive the process. Prompts 3 and 4 open this line of investigation and lead to prompt 5, the management agenda.

Prompt 3: *In order for the visible bits of this process to work, all sorts of other activities must be going on "behind the scenes". What else is being managed which you cannot actually see?*

Guiding questions:

- What systems are present in practice which form the essential means of running the process? What evidence is there of managing the operation through such systems as:
  - the order processing/appointments system which converts incoming requests into bookings for service
  - the design and specification system which converts the incoming requests into process instructions for action
  - the quality system which includes the recording of faults and the way in which the process is improving through the continuous learning of staff
  - the cost assessment system which provides estimates of the cost-of-service delivery
  - the job design and motivational system which provides staff with the basis for a satisfactory work experience
  - the coordination and planning system which prepares schedules and rosters for doing work, sourcing supplies or responding to constraints

- the maintenance system for ensuring the best uptime for the equipment.

Prompt 4: *Comment on the environment for the work.*

Guiding questions:

- What kinds of influence do you think people working in the process feel they have on their working conditions? Notice the speed at which people move, the evidence of supervision and the attractiveness of the work environment. In effect, ask what factors keep this process going and people trying.

Prompt 5: *The management agenda*

Most processes are arranged as a result of many incremental changes made over a long time. There is no reason to suppose that the resulting process is particularly good, or particularly bad, and it is worth exploring if the visible and "behind-the-scene" aspects of the process can be changed despite their apparent permanence on first appearance.

Towards developing the management agenda, prompts 1 and 2 are essentially analytical perspectives. Together they imply that the original design of the working process is potentially well-organised, i.e., that the overall scheme has an implicit design which considers flows, conversions and costs. However, the reality of running the process in practice is about coping with an endless set of puzzles and problems. Prompts 3 and 4 ask you to use your experience to try to envisage what kind of problems (or opportunities) arise and must be solved to keep the process working.

Understanding these four perspectives is central to the creation of the management agenda:

- How to change the capacity (resources made available over time) and to match it with the demand for service delivery.
- How to look for standardisation opportunities within the physical process while providing customisation required for individual service requests.
- How to introduce new services and associated processes into the current pattern of established working which itself may have become efficient in its current context.
- How to anticipate the impact of current resourcing pressures and new service opportunities on the process and its enabling systems.

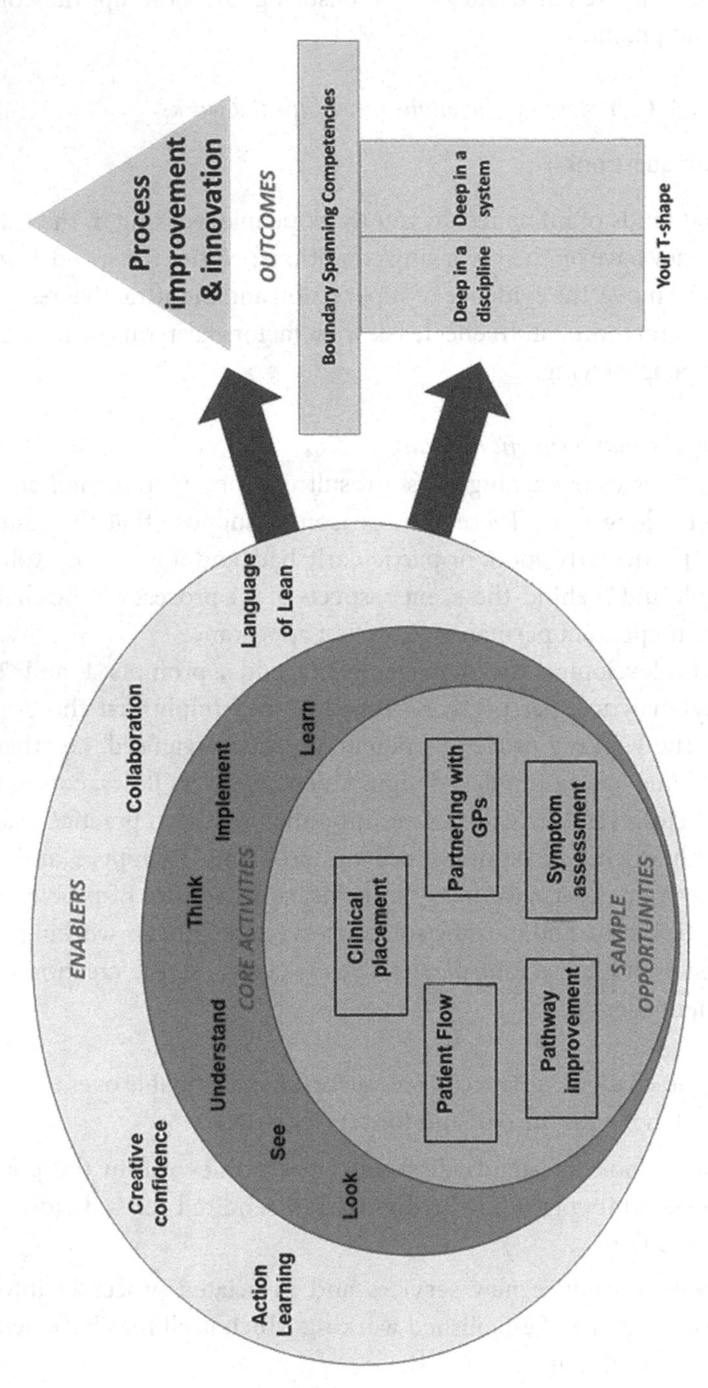

FIGURE 4.1 Linking and locating process improvement initiatives.

4.5 FROM PRACTICE

The approach outlined in Section 4.4 has encouraged and supported over 50 healthcare professionals to engage in understanding the multifaceted problems and challenges they face in practice and setting about solving them. Some examples follow, as illustrated in Figure 4.1.

The figure locates five process improvement and innovation initiatives undertaken by healthcare professionals. Table 4.1 summarises details of each of these initiatives.

Each initiative was developed and led by a healthcare professional following the approach outlined in Section 4.4. The development and implementation of the initiatives required that they looked at the processes for which they had responsibility. Here, they saw opportunities and came to understand the potential for improvement. Working with colleagues, they thought their ways through what it would take to make the changes before implementing their ideas. Learning emerged through the action of implementation and informed further improvements or replication in other locations. So, action learning combined with creative confidence, collaboration and the acquisition of a new process improvement language enabled the focal processes to improve. In addition, the healthcare professionals

TABLE 4.1 Process Improvement Initiatives

Focal Process	Target Improvement	Lead Role
Patient flow	To improve patient flow to and from the Radiology department	Radiographer
Partnering with General Practitioners	To deliver a service where the need of the patient is categorised and supported in a timely manner	Psychologist
Using tests as part of a screening pathway	To test all arriving passengers travelling from high-risk countries in the most efficient way possible to ensure limited entry of Variants of Concern into the country, while preventing a major bottleneck in the airport	Doctor
Respiratory symptom assessment and clinic management in primary care	Develop a process where the Respiratory Integrated Care team can offer post-Covid-19 respiratory assessment and symptom management clinics	Respiratory nurse
Clinical placement of Occupational Therapy students	Improve the induction process	Occupational therapist
The day surgery pathway	To enhance flow efficiency	Anaesthetist

themselves grew and developed their capabilities. They changed their T-shapes (Moghaddam, Bess, Demirkan & Spohrer, 2016), deepening their abilities to shape the system within which they worked and extending their competency in spanning boundaries to address problems with and through others.

4.6 CONCLUSION

Healthcare professionals face multifaceted problems and challenges and need to have the capability and confidence to set about solving them. However, like any condition in healthcare, if the characterisation is wrong, then the diagnosis may be incorrect and the solution inappropriate. In this chapter, we have explored the term "problem" and asked when a "problem" (colloquially) is not a problem (categorically); when is it a puzzle? It is not that problems require different prioritisation to puzzles. Rather, it is to recognise the difference and manage the approach to finding a solution differently. In particular, it is recognising the roles and engagement by the different stakeholders in achieving a solution. Collaboration, design thinking, creativity and learning are necessary. The context and mindset required to engage in this way feature in later chapters.

SOURCES AND SUGGESTED FURTHER READING

Coughlan, P. & Coghlan, D. 2011. *Collaborative strategic improvement through network action learning: The path to sustainability*, Cheltenham: Edward Elgar.

Moghaddam, Y., Bess, C., Demirkan, H. & Spohrer, J. 2016, T-shaped: The new breed of IT professional, *Executive Update*, 17: 8 (www.cutter.com).

Revans, R.W. 1998. *ABC of action learning*, London: Lemos and Crane.

CHAPTER 5

Understanding Creative Confidence, Design and Innovation

5.1 INNOVATION FOR IMPACT IN HEALTHCARE

The World Health Organization defines health innovation as a new or improved solution with the transformative ability to accelerate positive health impact (WHO website 2023). More generally, innovation is a practical process that includes the technical, design, delivery and management activities involved in making a new product or service available to users. So, innovation is more than an outcome or introducing new technology; it is a process which integrates ideas to add value for users in a product, a service or a process.

The word innovation is pervasive throughout healthcare. Indeed, many research departments associated with medicine and medical schools have extended their title to *Research and Innovation*, sometimes without thinking through what the scope of the title might mean. With regard to research, it can be defined as the enquiry into medical and scientific problems and the use of experimentation and studies to find answers. However, these answers are not in themselves innovations. Some key enablers of innovation which can dramatically improve work practice and healthcare service delivery include:

DOI: 10.1201/9781003123910-5

1. Medical technology, particularly medical devices and pharmaceuticals.
2. Information and communications technology from sophisticated all-hospital patient information systems to smartphone apps.
3. Systems and processes to use technology efficiently and appropriately. In hospitals, this area tends to be particularly weak and requires the most development.

In practice, the key opportunities for innovation in healthcare draw upon all three of the abovementioned enablers, and realising the potential for better healthcare and a better experience for the patient and the users requires that all three work in harmony. Patients have a right to the best technology, devices, medicines, methods and processes (all of which are the outcomes of the innovation process). Only by designing these outcomes in a manner that humanises solutions to healthcare problems and implementing outcomes that are widely used and have valuable impact in healthcare or medicine do they become innovations.

5.2 DESIGNING TO HUMANISE SOLUTIONS TO HEALTHCARE PROBLEMS

Design has a role in everyday problem-solving in the healthcare environment. It needs to be everyone's responsibility and, so, developing skills in this area is key. In healthcare, our design efforts while focused primarily on the patient need also to consider the healthcare staff who operate the system and others such as patient's families and friends who come into contact with it. In healthcare, design is less about products and more about processes – designing processes that facilitate healthcare professionals delivering responses to meet the needs of patients. During the course of experience in hospital, a patient will be admitted and then experience a diagnosis, hopefully followed by a treatment. During and following the treatment they will be monitored and, eventually, discharged, all the while experiencing appropriate and timely care. Said differently, the patient, as a user, should have a good user experience (UX) including an accurate diagnosis, appropriate therapy and the best of care (Figure 5.1). It is here that designing to humanise solutions to healthcare problems needs to include the perspective of the user.

Design can be a fascinating topic and, for design professionals, it can provide a challenging and rewarding career. It is a course of action for the development of an artefact. To produce successful products, designers often

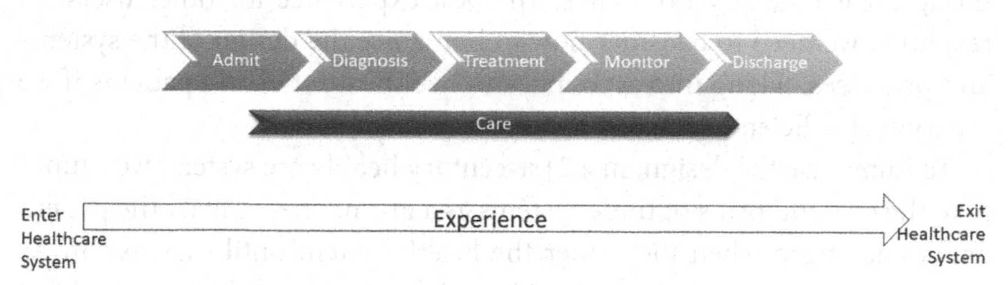

FIGURE 5.1 The patient experience.

need to consider the needs of users and draw upon multiple perspectives and viewpoints. In healthcare, we try to obtain these viewpoints by engaging many different stakeholders from disciplines and from the patient population. Some will argue that the product must be usable and understandable, others that it must be attractive and yet others that it must be affordable. Often, though, we are told that design is a very specialist area reserved for the few who trained and practiced in the different methodologies. For the finer points of design this is certainly the case. We relate it to the form and function of products like mobile phones, electric cars or even newer technologies like the electric taxi helicopter. However, design activity pervades organizations and as described by Peter Gorb and Angela Dumas (1987), 'silent design'. (that is, design by people who are not designers and are not aware that they are participating in design activity) goes on in all organisations.

Design is not a term we naturally associate with the care and management of patients in the healthcare system. Healthcare professionals aim to give patients the best diagnosis, the best treatment and the best care. These aims are reflected in the expression "providing the best quality of care". However, as technology infiltrates our lives, the lines blur: between the doctors, nurses and other healthcare professionals who provide the service; between the technology they use both in terms of medical devices and in terms of information and communications technology; and the processes that they use.

Many healthcare professionals are not particularly familiar with the potential of design. However, research in the area of design tells us that when you do not focus on the user experience, you will never get an optimum result. For example, Gambardella, Raasch and von Hippel (2017) explain when and how engaging with innovating users increases social outcomes. In healthcare, the needs of the doctor often dominate which, though relevant, may not deliver the best experience for other users. In response, we need to consider how to humanise the design of the systems and processes, taking into account both healthcare staff and patients if we are to work efficiently and optimise patient care.

To humanise the design, in a 21st-century healthcare system, we cannot take the current forms of these systems as a given. They impact the patient experience from when they enter the health system until they exit it. As such, the design of solutions should not be just based solely on specialist knowledge of healthcare staff but also the experience of the users of the system, process, technology or device. Identifying and understanding this experience requires an empathetic approach and new skills and ways of working for healthcare professionals.

If we are to design and develop a system that considers all these factors, we need to invest time in understanding or empathising with the group that we are designing for. As noted earlier, even where the primary user is the patient, other users need to be considered too, such as the patient's family, the healthcare professionals and others who work in the setting. Further, the ability of patients to develop new medical products to serve their own needs is growing and merits attention, with the promise of application in medically and socially valuable directions (Demonaco, Oliveira, Torrance, Von Hippel & Von Hippel, 2019).

5.3 IDEATION AND PROTOTYPING

The act of design does not in itself mean creativity or that the solution will work in practice. Health facilities and hospitals are the settings in which we find the healthcare workplace. This workplace is very much ruled by standards, known practices, processes and policies, adherence to which ensures quality of care. However, it is logical to assume that, in this setting, new ideas can be disruptive. So, if we are to develop new ideas about work practices or healthcare delivery, we must find a way to utilise our collective creativity while reducing risk.

Often in the modern workplace, we can be asked to generate and contribute our ideas. Idea generation is any activity we partake in that results in the creating, developing or sharing of new ideas, be they by abstract, virtual or physical methods. The term "ideation" is often used when we express those ideas by sharing them with others. Rapid prototyping is a fast, efficient and inexpensive way to share and express ideas, usually in a visual way and often in a physical way as well.

In healthcare, however, trialling ideas that have a direct impact on patient care can be risky. Experimental changes to a patient-related process could result in delays in care and could easily be considered unethical or worse, causing inaccuracies or errors. So, while testing and trialling solutions to problems may not be possible in the live environment where patients are cared for, it is possible to simulate and prototype ideas and discover their strengths and weaknesses before deciding whether or not to put them into practice. Prototyping also has advantages in a healthcare environment, where with prudent thinking and technique, you can surface and "iron out" potential problems in a proposed new process before actually bringing it into practice.

Prototypes have many purposes. Ulrich and Eppinger (2012) describe a prototype as being able to facilitate integration, communication, learning and milestone management. This range of purposes enables designers, from the outset, to consider the particular purposes of the prototype in prospect, and whether one or more may be useful. The different types of prototype may be introduced at different phases of the innovation initiative demonstrate proof of concept, proof of product, proof of process and proof of production or delivery.

Prototypes may be represented virtually or physically in different materials, shapes and sizes. They can demonstrate a particular component or the comprehensive product or service. One way of distinguishing among them is to imagine prototypes on a spectrum of high-fidelity (hi-fi) to low-fidelity (lo-fi), where the fidelity of the prototype refers to how closely it resembles or how similar it is to the planned product (Coutts, Wodehouse & Robertson, 2019). Lo-fi prototyping is characterised by quick and easy creation of an artefact. It can be low-tech and realised in cardboard, sticky tape or post-it notes. Low-fi prototyping can be low cost and open to many stakeholders to contribute to the design discourse. In contrast, the hi-fi prototype may use the same materials and production/delivery methods as the final product or service in prospect. As such, it can demonstrate more accurately the promised functionality and appearance. Hi-fi prototypes take more time, money and resources to make than for lo-fi prototypes. As to which is better, it depends on the objectives and on how much time and money are available. For example, Figure 5.2 shows a lo-fi prototype built in a few hours using the LEGO® Serious Play Method during a workshop to explore strategy with a leadership team at a large healthcare organisation.

FIGURE 5.2 A low-fi prototype.

5.4 WHAT IS CREATIVE CONFIDENCE?

So, with a sense of innovation, design, ideation and prototyping, what is there to stop the healthcare professional from being creative and innovating? Simply put, the answer may be fear. It is said that when we are born, we are naturally creative. Indeed, most of us can recall times when, as children, we played imaginary games, drew imaginary characters, created a play space from what was immediately to hand or otherwise displayed creativity. Innovation practitioners like Tom and David Kelley have built their careers and reputations by connecting people with this innate childhood creativity (Kelley & Kelley, 2012).

Despite this sense of a creative birth right, we can find that the structured worlds of education, training and work can restrict our opportunity to be creative and replace it with an emergence of caution. Figure 5.3 visualises this progression: as we age and mature, we become more experienced through our lives and our work, but we also become more cautious and more analytical. So, for many people, it is their very work and life experiences that stifle their creativity and, so, their thinking becomes fixed or stuck. Despite this, there are ways to recover that creativity through rebuilding the creative confidence that we had when we were children or young adults.

This challenge is relevant. The opportunity and the need for creativity remain and an ability to think creatively is at the very core of how we solve

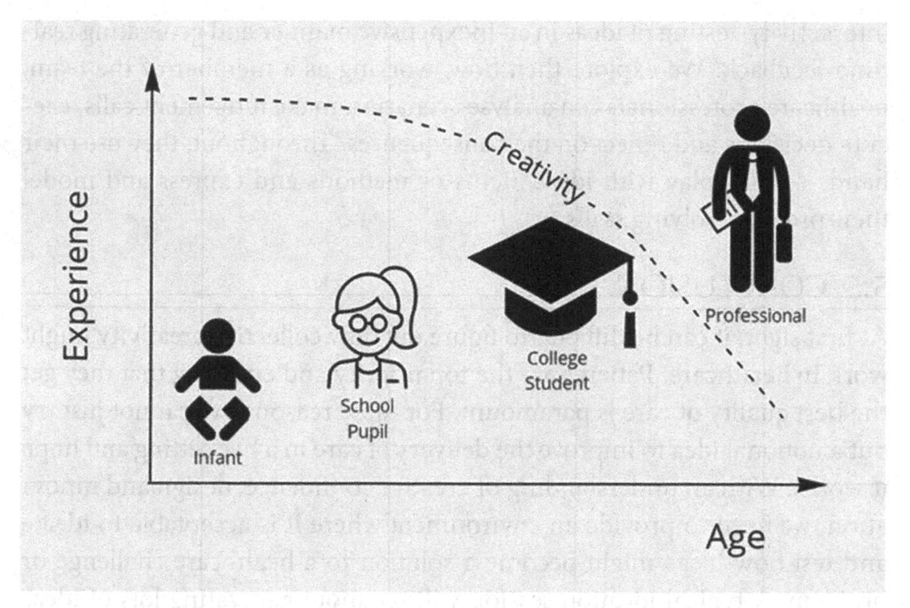

FIGURE 5.3 Creativity, age and experience.

multifaceted problems in the modern sophisticated world of healthcare. The networked wider world we live in, driven by the digital economy, has learnt how to exploit human creativity and much of what is learnt can readily be transferred into healthcare. But, how do we overcome the caution and the constraints? Fundamentally, being creative means to have a mindset and approach open to developing new ideas. However, ideas are not solutions, and we need to have lots of new ideas if we are to find novel solutions to the problems and challenges we face. To realise the potential of those ideas, we have to overcome our fears. We described the four fears identified by Kelley and Kelley (2012) earlier in Chapter 2 when sharing the case titled *Be fearless and innovate*: fear of the messy unknown, fear of being judged, fear of the first step and fear of losing control. Creative confidence arises from overcoming these fears.

A useful and novel method to encourage healthcare workers and professionals to overcome their fears and become more creative is to use play methods. Play can help people break out of their fixed thinking patterns and allow them the freedom to generate much more creative solutions to the multifaceted problems found in healthcare. In Chapter 7, we explore play and gamification as a way of allowing individuals to engage with ideas in action by creating an environment which incorporates interactivity, testing of ideas in an inexpensive manner and generating real-time feedback. We explore then how, working as a member of the team, healthcare professionals can analyse scenarios, make judgement calls, execute decisions and reflect on the consequences. Throughout, they use their hands as they play with ideas, items or methods and express and model their problem-solving skills.

5.5 CONCLUSION

At first sight, it can be difficult to figure out how collective creativity might work in healthcare. Patients are the top priority and ensuring that they get the best quality of care is paramount. For these reasons, we cannot just try out a notional idea to improve the delivery of care in a live setting and hope it works. With an understanding of creative confidence, design and innovation, we need to provide an environment where it is acceptable to ideate and test how ideas might become a solution to a healthcare challenge or problem. A typical ideation session will set about generating lots of ideas and use workshopping methods to filter and parse these ideas. However, these methods of working are not innate to the healthcare setting and can be defined as new ways of working. For this reason, they must be mastered or, more importantly, experienced and practiced.

We generally refer to this mastery as upskilling since the staff involved in these activities are otherwise highly skilled and trained in their chosen speciality and, so, are adding in a significant way to their skill set. A novel and useful method to encourage healthcare workers and professionals to become more creative is to use play methods. Play can help people break out of their fixed thinking patterns and allow them the freedom to generate much more creative solutions to the multifaceted problems found in healthcare. We develop these themes further in Chapter 7.

SOURCES AND SUGGESTED FURTHER READING

Coutts, E.R., Wodehouse, A. & Robertson, J. 2019. A comparison of contemporary prototyping methods, Paper presented at the *International Conference on Engineering Design*, ICED19, August 5–8, 2019, Delft, The Netherlands.

Demonaco, H., Oliveira, P., Torrance, A., Von Hippel, C. & Von Hippel, E. 2019. When patients become innovators. *MIT Sloan Management Review*, 60 (3), (Spring 2019), 81–88.

Gambardella, A., Raasch, C. & von Hippel, E. 2017. The user innovation paradigm: Impacts on markets and welfare. *Management Science*, 63 (5), 1450–1468.

Gorb, P. & Dumas, A. 1987. Silent design, *Design Studies*, 8 (3), 150–156.

Kelley, T. & Kelley, D. 2012. Reclaim your creative confidence. *Harvard Business Review*. December, 115–118.

Ulrich, K. & Eppinger, S. 2012. *Product design and development*, 5th ed, Boston, MA: Irwin/McGraw-Hill.

WHO website. The WHO Innovation Scaling Framework: www.who.int/teams/digital-health-and-innovation/health-innovation-for-impact

CHAPTER 6

Where Do Mindset and Climate Fit?

I wanted ... people that were ... innately curious and people that had ... a sense of humour about themselves, because we will probably be making a lot of mistakes. So 'are they people who can ... roll with it and laugh at themselves and be OK with it?'

(Carlgren, Rauth & Elmquist, 2016, p. 48)

6.1 MINDSET

The idea of mindset as a way of individual embodiment of principles is central to Design Thinking (DT). Carlgren, Rauth and Elmquist (2016) conceptualise DT as consisting of a number of core themes that are embodied and enacted as a set of principles/mindset, practices and techniques. In particular, they suggest that the focus on specific ways of thinking, attitudes and cognition emphasises the importance of individuals and the ways they interact in DT. Table 6.1 links some characteristics of mindset to DT.

Mindset impacts how DT is addressed in organisations, what actions might be planned, and how DT might be evaluated. For example, in order to empathise a mindset might be characterised by openness, avoiding being judgemental and being comfortable around people with different backgrounds and opinions. Similarly, a mindset supporting experimentation

DOI: 10.1201/9781003123910-6

TABLE 6.1 Linking Mindset and Design

DT Themes	Mindset
User focus	• Empathic • Curious • Non-judgemental • Social
Problem framing	• Unconstrained thinking • Comfortable with complexity and ambiguity • Open to the unexpected
Visualisation	• Thinking through doing • Bias towards action
Experimentation	• Curious and creative • Playful and humoristic • Optimistic and energetic • Learning oriented • Eager to share

Source: Based on Carlgren, Rauth and Elmquist (2016).

might be characterised by curiosity, playfulness, optimism and displaying a sense of humour. Further, creativity and unconstrained thinking might characterise a mindset geared towards action.

Carlgren, Rauth and Elmquist (2016) observed further that the DT approach has the potential to fundamentally shift mindset about how to relate to user knowledge:

> *... a crucial part of DT entails gaining a thorough understanding of users and their needs, even those they are unaware of. The practices employed to arrive at this understanding are generally carried out through ... qualitative research, asking what really motivates [the users], what their purposes are and what they do.*
>
> *(p. 46)*

Of course, there is no "right" mindset. If anything, diversity of capabilities, hierarchy and mindset can strengthen a team, provided that it is underpinned by a democratic spirit, and openness to differences in backgrounds.

The transition to a new mindset and the establishment of new routines are complex and difficult, especially when mindset may contribute to barriers to the adoption of DT. Sandberg and Aarikka-Stenroos (2014) divide barriers to innovation into those internal or external to the organisation. They identified a restrictive mindset, one particular internal barrier. Assink (2006) proposed that mindset barriers at individual and organisational levels were internal inhibitors to disruptive innovation. So, individuals with an analytical mindset might inhibit innovation in contrast to those with an interpretative mindset, tolerant of ambiguity and willing to renew competencies. Similarly, frontline healthcare staff may be risk-averse as a natural part of their professional role. As such, a demand for rapid testing of hypotheses or failing and learning from mistakes, as associated with DT, may be difficult in healthcare organisations with a defensibly risk-averse culture.

6.2 LEARNING AND CLIMATE

Creativity in organisations relies on a creative climate. Basically, climate refers to a set of attitudes, feelings and types of behaviour that emerge on a daily and collective basis within the organisational environment. Correspondingly, a creative climate is embedded within the organisational work environment, and both promote and allow a collective expression of different viewpoints. This behaviour adds value and enhances creativity within the organisation. The Creative Problem Solving Group (1992) developed one of the most widely used models for thinking about creative climate. The elements are outlined in Table 6.2.

Expressed in this way, the climate supports the opportunity for creativity and learning, generating new ideas and communicates them to other stakeholders. It encourages interaction, critique and design improvements to processes, systems and artefacts. When embedded within the broader healthcare environment, it follows that there can be positive motivational consequences and agency for staff arising from exploring and exploiting their creativity.

Amabile (1997) proposed a framework which identifies five environmental components that impact individual creativity:

1. encouragement of creativity;
2. autonomy or freedom;
3. resources;
4. pressures;
5. organisational obstacles.

Here, encouragement from the organisation and/or supervisors, workgroup support, adequate resources and challenging work combine to stimulate creativity. In contrast, organisational obstacles (such as internal politics,

TABLE 6.2 Creative Climate

Element	Indicator	For Example
Challenge	The degree to which members of the organisation are involved in its daily operations and in setting long-term goals.	In a challenging organisation, the members are intrinsically motivated to make contributions to the organisation's success
Freedom	The ability to exert independent behaviour in the organisation.	Organisations with high degrees of freedom allow their people to autonomously define much work.
Trust/ Openness	The emotional safety in relationships.	In an organisation characterised by a high degree of trust, people generally feel free and comfortable in putting new ideas forward.
Idea time	The amount of time people can and do use for elaborating and propounding new ideas.	In a high idea situation, possibilities exist to discuss and test new, spontaneous suggestions
Dynamism/ liveliness		The organization is an exciting place in which to work
Playfulness/ humour	A relaxed, spontaneous atmosphere characterised by joking and laughter.	Joking in a good-natured way to enjoy the work
Idea support	The manner in which the organisation treats new ideas.	In a supportive climate, ideas are received in a kind and attentive way. Members listen to each other and encourage each other's ideas
Debate	Clashes and arguments over viewpoints and ideas.	In an atmosphere of debate, many people's ideas are heard and discussed
Conflict	Personal and emotional tensions in the organisation (as opposed to idea tensions in the debated dimension).	Conflict is contrary to creativity and is to be constructively addressed and quickly resolved
Risk-taking	The tolerance of uncertainty and ambiguity in the organisation. In an organisation that takes risks, initiatives can be taken even when the outcomes are uncertain.	Members feel that they can take a chance on new initiatives and put ideas into action.

Source: Based upon Creative Problem Solving Group (1992).

harsh criticism of new ideas, destructive internal competition, avoidance of risk and overemphasis on the status quo) and workload pressure (such as extreme time pressure and unrealistic expectations for productivity) hinder it.

The highly interactive behaviour of staff is a feature of an effective climate. Trust is placed in the staff members and encourages them to take risks to create novel designs and concepts. In order for the organisation not to sink into chaos, however, an effective climate must strongly emphasise idea sharing and have respect for a diversity of opinions among all work colleagues. In this setting and during sessions for problem-solving and ideation, staff members must become comfortable with uncertainty, incomplete information and paradox. There should be a tolerance for mistakes and an atmosphere that encourages experimentation. In this context, psychological safety must permit staff to feel safe at work in order to grow, learn, contribute effectively and perform. If the right climate is created, staff can be self-starting, proactive, persistent in surmounting obstacles to achieve a goal, and orient their efforts towards the long-term goals of the healthcare organisation and improving care for their patients. Here, the established relationship between psychological safety and knowledge sharing behaviour can be strengthened when self-starting and persistence in implementing goals are recognised and valued.

Finally, Cirella et al. (2016) noted that creativity is a process based on learning, which can be planned, institutionalised, formalised and designed with cognitive, structural and procedural learning mechanisms. It is not only about "creative individuals" but is an organisational competence that can be improved upon or hindered by these organisational learning mechanisms. In general, such practices and actions aimed at learning are associated with a more creative climate. In particular, the cognitive learning mechanisms link directly to mindset.

6.3 INTEGRATING MINDSET, LEARNING AND CLIMATE

At a corporate level, traditional hierarchical culture obstructs creative thinking and innovation. Managers and senior executives can be reluctant to express disruptive ideas for fear of judgement. Yet, creativity is needed to stimulate innovation to improve practice. Many ideas can come from all levels and areas of the workforce. In particular, younger people often have novel ideas that challenge established hierarchy. In like manner, large healthcare organisations such as a hospital or hospital group have the potential to stifle innovation. Not only do they have an administrative hierarchy but also a medical/professional hierarchy and, sometimes, these hierarchies are not working towards shared goals. In addition, the core team of senior executives is often caught up in BAU (business-as-usual) and does not have the time, headspace or focus to immerse themselves in innovative practice. They may even see such practice as a distraction from running around and getting the busy work done.

Yet, there is a way forward. Small- and medium-size companies often are run by a core team, which tends to allow for a more collaborative and innovative approach. Staff have to roll up their sleeves more and become familiar with joint roles or covering different roles when required. This flexibility is based upon an understanding of how the overall organisation runs and facilitates the development of empathy that supports innovation. In contrast, in healthcare, both training and practice are highly specialised and sharply focused when deployed. As such, the professionals can have less empathy with other areas and a mindset oriented, at best, to innovation in their departments, groupings or specialities but far from the overall goals of the healthcare organisation. However, this mindset does not necessarily support the wider organisation to run smoother and more effectively. In fact, that mindset can even preclude staff from getting involved in larger organisation-wide problem-solving and display the "nothing to do with me and my department" attitude.

6.4 HOW CAN WE CHANGE OR INFLUENCE MINDSET AND CLIMATE?

For healthcare practitioners, the multidisciplinary team is a common term for a range of disciplines coming together to solve difficult patient cases. However, rather than facilitating a meeting of minds towards an interdisciplinary action, it can become a soapbox for disciplines and clinicians involved in a patient's care to stand up and declare on the primacy of their therapeutic and diagnostic findings. Such a climate is not necessarily conducive to interaction and innovation.

The work structure for healthcare professionals, based on their educational systems and their training practices, can lead them to focus largely on problems and puzzles within their own specialist field. However, as problems become more multifaceted, these workers need to reflect on a range of aspects to care delivery that might impact their ability to contribute to solutions. While individual healthcare professionals have a mindset where they care, it can be difficult for them to be innovative in a climate with rigid and traditional delivery systems. Further, healthcare professionals who are over-regulated and, often, over-worked can slip into a fixed mindset. They try to avoid challenges that fall outside their role definitions, can find it difficult to overcome obstacles to delivering their specialist service and are not appreciative of feedback which they find challenging. In effect, they are constrained by a system which may not have been designed with the patient experience at its core. This situation continues to exist but needs to change if healthcare is to take advantage of the efficiencies and effectiveness that advances and overlaps in the digital and medical technologies can bring. As disruptive innovations in their own right, these advances are shaking the foundations of how medicine, medical care and healthcare delivery are presented to and experienced by healthcare professionals and patients. It is here that mindset and culture surface again.

Start-up culture, particularly the type that has developed in the last 25 years, around medical devices and ICT (information and communication technology) can teach us a lot. Often, it is a small number of collaborators with diverse skills but focused on a common mission with passion and vision that achieve the greatest success. These start-up organisations, and the innovators that run them, exemplify what could work in healthcare delivery to make work practices, processes and systems more efficient and integrated. The innovators understand that success and work effectiveness are no longer driven by the organisation structure, power and scale but more by imagination and creative confidence. Work practices in many fields are driven less by manpower and more driven by mindpower. They tend not to be led by one leader but more by a leadership team drawn from diverse areas of responsibility across the organisation.

Of course, it can be difficult for those working in a high-pressure healthcare setting to step out of their environment, become more entrepreneurial (more appropriately intrapreneurial), and develop and use a growth mindset. Yet, it is a contention of this book that healthcare professionals and the system within which they practice can benefit from developing such a growth mindset. This mindset can allow them to enhance their wider and more transferable skills so that they can embrace wider challenges, deal with setbacks and learn from critique. These professionals can be inspired by others both from within and outside their professional sphere. Said differently, a growth mindset can allow them to develop their T-shapes, discussed earlier in Chapter 4 (Moghaddam et al., 2016). T-shaped professionals typically have depth of skill in at least one discipline and one service delivery system. Together, they represent the vertical leg of the T. These professionals are also broad in that they can collaborate effectively with others who are deep in other areas so that they can innovate, co-create and solve complex problems.

Hospitals and other healthcare institutions need then to become settings where doctors, nurses and other healthcare professionals can question how they diagnose and respond to challenges and problems with a different mindset. They recognise the value of collaboration with colleagues in diverse teams, so as to identify novel solutions which, together, they can implement and learn from. To achieve such a change, we need to develop the mindset of key workers at all levels in the organisation, the climate in which they work, their skill set and give them the confidence to continually question how they can improve, and problem-solve continuously.

6.5 CONCLUSION

Healthcare professionals face particular challenges and problems. Their mindset and climate influence how they approach them. Figure 6.1 visualises how healthcare workers might deal traditionally with problems and challenges (in black) and how they might behave creatively (in red). In essence, it illustrates the classic dilemma between fixed and growth mindsets. During service delivery when there are problems, the instinct is to try to fix the problem and return things to the status quo (black line on time versus effort graph). But in reality, we are continuously facing problems and challenges. So, to be good problem-solvers we need to grow in a work environment where we are open to problem-solving. Said differently, we need to develop the confidence to continuously discover and question how to solve ongoing problems and challenges by using innovation and creative thinking.

Towards that end, we recognise that people who have challenging performance goals can be worried about how they are perceived and, so, avoid

Challenges and Problem Solving In Healthcare

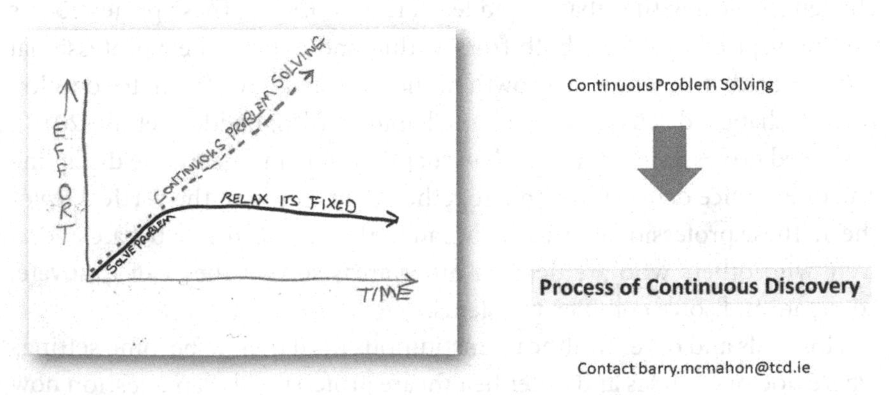

FIGURE 6.1 Challenges and problem-solving in healthcare.

work that exposes them or takes them out of their comfort zone. Most healthcare workers have a deep knowledge in their specialist field or discipline but if they are to be effective in interdisciplinary innovative activities, it is essential for them to develop such transferrable skills as teamwork, communications, networking, critical thinking, global understanding and project management. These transferrable skills can extend their T-shape and prepare staff to integrate and perform as part of an interdisciplinary team working to solve key healthcare challenges.

We develop this theme further and, in Chapter 7, we describe how using playfulness among healthcare professionals can improve the climate and change the mindset. We emphasise that it is not play for its own sake. Rather, play is a vehicle for education on ideation, design and innovation. Just as in the wider organisation, playfulness requires its own climate. We describe how to create that space for creativity and innovation.

SOURCES AND SUGGESTED FURTHER READING

Amabile, T.M. 1997. Motivating creativity in organizations: On doing what you love and loving what you do, *California Management Review*, 40 (1), 39–58.

Assink, M. 2006. Inhibitors of disruptive innovation capability: A conceptual model, *European Journal of Innovation Management*, 9, 215–233.

Carlgren, L., Rauth, I. & Elmquist, M. 2016. Framing design thinking: The concept in idea and enactment, *Creativity and Innovation Management*, 25 (1), 38–57.

Cirella, S., Canterino, F., Guerci, M. & Shani, A.B. (Rami). 2016. Organizational learning mechanisms and creative climate: Insights from an Italian fashion design company, *Creativity and Innovation Management*, 25 (2), 211–222.

Creative Problem Solving Group. 1992. *Climate for Innovation Questionnaire*, Creative Problem Solving Group, Buffalo, NY.

Moghaddam, Y., Bess, C., Demirkan, H. & Spohrer, J. 2016. T-shaped: The new breed of IT professional, *Executive Update*, 17 (8) (www.cutter.com).

Sandberg, B. & Aarikka-Stenroos, L. 2014. What makes it so difficult? A systematic review on barriers to radical innovation, *Industrial Marketing Management*, 43, 1293–1305.

CHAPTER 7

Creating a Climate for Creative Thinking, Design and Innovation

7.1 TOWARDS A CREATIVE APPROACH TO CHALLENGING AND PROBLEM-SOLVING

Some of the key tenets of using creative thinking involve how to lead and mentor key staff (Jobst et al., 2012). This practice puts a clear emphasis on action and education. Our contention in this book is that a key part of steering healthcare staff towards a creative approach to challenge and problem-solving is to get them to practice the art of generating novel and competitive ideas that have the potential to be useful (Cirella et al., 2016). To develop this theme, we break this effort into staff development and upskilling through action learning.

7.1.1 Upskilling

As we have noted throughout, healthcare professionals have already studied and trained to high clinical or technical levels. Most are motivated and dedicated to their work and to the patients they care for. However, for creativity and innovation to occur, it is not just about individual dedication and willingness but also how to look for and see opportunities in practice. For many, the associated questioning, reflection and response demand upskilling which complements their core disciplines. One of the

DOI: 10.1201/9781003123910-7

best methodologies for learning and upskilling is exposure to opportunities in practice and engagement in problem-solving through experiential or action-based learning. Schein (2014) saw experiential learning as creating the conditions which force learners to draw upon their inner cognitive and motivational resources. For example, skills can be developed through active collaboration, and in training workshops which bring creative thinking and tools to bear on real-world problem-solving. The techniques used and developed involve design thinking, game storming and serious play principles. This experience animates the learners and can produce not only more intense and stable learning but can teach the learner how to learn (Powell & Coughlan, 2020).

7.1.2 Action Learning

Action learning is an approach to learning in action where individuals with a commitment to action and learning form a group to address a problem for which there is no single or technical solution (Coughlan & Coghlan, 2011). Together, they engage in processes of questioning and reflecting on actions taken with a commitment to learning as candidate solutions emerge and are implemented. As a learning and coordinating mechanism (Coughlan & Coghlan, 2011), action learning can be deployed at different levels in an organisation, such as senior management, healthcare delivery unit or team. This deployment can involve designing action learning practices into process improvement initiatives and training events within departments, as well as into the management meetings. Here, there is an opportunity for facilitators to act as action learning coaches, breaking people out of their normal or traditional environment. A number of design choices are available to facilitators in order to support this action learning:

Space – Physical space helps to create (or inhibit) a safe space for creativity. It helps if there is a special or dedicated room where the training and workshops are carried out. In a healthcare facility, a meeting space with re-configurable furniture, lots of whiteboards, access to Wi-Fi and coffee can work. The idea is to facilitate staff in identifying an appreciative atmosphere, free from the normal constraints of their routine work. It should not become a space to talk about routine job responsibilities, such as patient cases, or to have traditional management meetings.

Away-Day – For special work or topics or for teams that are sharply focused on the day-to-day (often referred in healthcare as business as usual (BAU)), it can be useful, particularly in the early stages of creating an environment for creative thinking, design and innovation, to take staff completely out of their normal environment and into a safe workspace for a day. These workspaces are often now available in co-working environments, incubation facilities and start-up communities.

7.2 LEVERAGING GAMES

A critical choice available to facilitators in order to support action learning is to play a game. So, here, we introduce the related concepts of gamification, playfulness and metaphors, the various uses of play tools and play in the healthcare setting.

7.2.1 Gamification

Gamification introduces game design ideas into non-game activities in order to make them more engaging and even fun. These ideas can include competition, points, achievement, rules of play, status and self-expression so as to encourage actions through positive feedback. While the fit of gamification in healthcare might seem alien to the serious challenge of delivering healthcare, particularly in the hospital setting, it is a clear and structured method to allow creativity to flow in diverse teams and to find imaginative and disruptive responses to complex problems. It is also a useful and usable method to link the structured delivery of healthcare with the messy and awkward process of creativity. While play is a universal human experience across cultures, it has also been the focus of scholarly inquiry from which theoretical and empirical evidence has underscored its significance in the process of individual expression and adaptation (Kolb & Kolb, 2010). Games can also provide spaces, tools and media for the application of creative thinking, reflection and learning.

The term "gamification" commonly refers to the use of game design methods as a means to leverage games for benefit. As a concept, gamification relates to games, but not to play or playfulness. There is a difference between a game and play which can be illustrated by reference to Caillois' (1961) concepts of *paidia* and *ludus* as two distinctive poles of play activities. While *paidia* (playing) refers to a freedom to choose and take voluntary actions, *ludus* (gaming) denotes a rule-based gaming process with well-defined sets of rules and regulations for objectives to be achieved. Caillois (1961, p. 12) identifies four ideal types of game:

- Games based upon competition
- Games incorporating chance
- Games of mimicry
- Games that expose the players to physical forces and movements that cause a sense of being released from the human body

These four ideal types of games can be combined into hybrids.

The use of play in the workplace can create new solutions in an exploratory, playful and imaginative way. Such use is crucial to creative thinking, ideation, innovation and developing future scenarios. The concept of play enables the linking of ideas and action and is the medium between thinking and doing. When playing with objects, participants can unlock their creative thinking through making and re-affirming the deep relationship between their hands and their brain (Wilson, 1999).

There are different kinds of play. In "frivolous play", play is used as pure amusement. In contrast, "serious play" has purpose and structure. The term "serious" is related to the objective; the expression "play" relates to the creative process (Schulz, Geithner, Woelfel & Krzywinski, 2015). So, serious play refers to an array of playful inquiry and innovation methods that serve as vehicles for complex problem-solving, typically in work-related contexts. LEGO® SERIOUS PLAY® is one of the best-known examples; however, serious play methods also include improvised theatre, role-play exercises, low-fidelity prototyping, as well as simulations. Serious play emphasises an exploratory mode that enables behaviour in which participants experience a sense of curiosity that can lead to a highly creative mental state. Overall, play contrasts with work (Styhre, 2008). Where work can be constrictive, structured, tedious, difficult and boring, play can be liberating, unstructured, refreshing and emotional. Playing implies rule-following while simultaneously moving beyond these rules. There is an element of play in all creative thinking, ideation and innovation effort.

7.2.2 Playfulness and Metaphors

As children, we played. It was enjoyable and we learned about ourselves and others. As adults, play can be more challenging. We may be sensitive to being judged by others and, so, may engage in play only with our children.

Their trust in us and respect for us gives us the licence to engage creatively. Yet, in the work life of many individuals, playfulness and acting/role-play has been shown to:

- Stimulate the brain toward creativity
- Help staff to relax in the work environment
- Encourage interactions with more diverse individuals, teams and groups
- Provide clear evidence that drawing on one's inner child and reconnecting with our creative confidence ultimately drives productivity and performance (Zenk, Hynek, Krawinkler, Peschl & Schreder, 2021)

Play brings its own language, and metaphor can be seen as an essential characteristic of the creativity of language (Ortony, 1993). Metaphor affords different ways of perceiving the world, sometimes in ways which are enabling or constraining. However, metaphor does not have to be the villain; the alternative ways of seeing that it affords not only an advantage in educational contexts but also a necessary feature of them.

Metaphor can play an important role in the formulation and transmission of new ideas through the language used in a serious game. Many problems in healthcare are multifaceted and can present as complex, difficult to resolve and challenging to communicate. As an ingredient of language acquisition, metaphor can make a very complex problem easier to understand and articulate. It can help a team to collaborate by creating a language for their shared vision of the challenge.

7.2.3 Use of Play Tools

Human evolution has demonstrated that the development of the human brain is directly linked with our evolution from a four-legged creature to a two-legged, two-handed one. Further, we think much deeper about a subject when we engage our hands (Zenk et al., 2021). Gaming and playing are obvious ways to use our hands while thinking. It makes sense that if we engage our hands in the process of thinking out a problem or dreaming up innovative ideas, we may just enrich that process.

7.2.4 Using Play in the Healthcare Setting

Many of the workshops that the authors run with healthcare professionals incorporate the principles of action learning while the creative thinking practices used are often closely aligned to play. We often say at the beginning of a workshop *"if you don't feel a little bit uncomfortable and a little child-like during the workshop, then we've done something wrong"*.

As noted earlier, the workplace setting in healthcare tends to be hierarchical and siloed. When there are problems and challenges, professional groups meet up to discuss but the environment can be formal and there is a distinct hierarchy. So, it can be difficult to marry the understandable seriousness of resolving problems in delivering healthcare for patients, with the seemingly casual and often unstructured nature of play. However, this nature can be the very reason why play can be such a contributor to creative thinking in healthcare.

Theming the play has a central role in connecting the activity to the healthcare delivery reality. In learning and teaching, we speak of learning objectives. In policy formulation, we speak of policy objectives. Such theming aims to create a holistic and cohesive direction, culture and organization. As a method, theming can articulate a narrative for interaction and a guide to structuring the delivery of objectives. In healthcare workshops, themes, such as those below, can guide the framing of objectives for serious play:

1. Responding and adapting to the constantly changing environment in which healthcare must be resourced and delivered.
2. Enhancing and improving the healthcare outcomes of our citizens.
3. Protecting current and future patients from the impacts of extreme acute and chronic healthcare events.
4. Developing a collective understanding of the medium- and long-term operational resilience and infrastructural challenges faced by users of the healthcare system (including staff, patients and their families), and finding solutions to mitigate them in sustainable and efficient ways.
5. Developing and testing new ways of working and implementing core activities so as to deliver greater value for patients and healthcare staff.
6. Exploring opportunities associated with new technological tools to stimulate innovation and new forms of collaboration among and between patients, their families and healthcare staff.
7. Encouraging new processes and operational models and service offerings that improve the hospital user experience, especially for those most vulnerable.

Actions to advance such themes demand new and different thinking, collaboration and confidence. The concepts and approaches presented in this book are candidates for application.

7.3 LEGO® SERIOUS PLAY®

Having introduced the related concepts of gamification, playfulness and metaphors, we explore now the use of LEGO® SERIOUS PLAY® in the healthcare setting. Nowhere is the principle of play more apparent than in the use of this methodology. It is a process which stems from the broad familiarity with LEGO® bricks and components and the flexibility of LEGO® building system. In practice, we have used it successfully as a tool to unlock the imagination of healthcare staff and to facilitate innovation within healthcare organisations. The idea of the LEGO® SERIOUS PLAY® methodology originated in 1996 when Professors Johan Roos and Bart Victor of IMD in Switzerland and LEGO® Group CEO and owner, Kjeld Kirk Kristiansen, were exploring alternative strategic planning tools and systems. They shared an appreciation of the value of employees and developed the concept of an inclusive, evolving and adaptive methodology to support creative thinking, design and innovation. The approach was named LEGO® SERIOUS PLAY® and it used LEGO® elements as a way of creating three-dimensional models of business issues and challenges.

7.3.1 Training in LEGO® SERIOUS PLAY®

The LEGO® SERIOUS PLAY® methodology is highly developed. It is therefore relatively straightforward to learn and apply. In Europe, particularly, there are trained instructors and experts providing training to a wide network of people from an array of industry sectors. It is still, however, relatively underdeveloped in healthcare. The methodology involves getting individual participants to express their particular understanding of a situation, ideas for improvement or other thinking in the form of a metaphor, represented in a LEGO® model. Through modifying the individual models and, in a team, looking at potential interconnections between the set of individual models, participants can build empathy with each other and the broader set of users, represented by a wider landscape of models. Finally, through identifying the priorities in the landscape, participants can develop a shared and uniquely creative vision for the particular challenge they set themselves.

7.3.2 Experiencing LEGO® SERIOUS PLAY®

Meetings are useful to deal with many issues. However, issues demanding creativity benefit from a workshop. Each format has a different approach to information exchange and interaction, as summarised in Figure 7.1. Our experience is that using LEGO® SERIOUS PLAY® requires a workshop and especially so for healthcare staff who tend to have particularly complex daily work routines. But, as Figure 7.1 points out, meetings can be effective in informing and exchanging ideas especially when everyone understands the terms of reference. However, if the issue merits "rolling up the sleeves" so that a group can develop a new plan of action on an important topic or reach a consensus on how to move a problem forward, then the workshop format is the better way to take action rather than talk about it.

The secret to a good workshop is understanding the starting and end points of the challenge and planning the journey in advance. Then, it is necessary to have someone available with facilitation skills to bring the team together. In the case of the workshops we run for healthcare professionals, as facilitators, we use LEGO® SERIOUS PLAY®. In advance of the workshop, we plan and coordinate arrangements with a leader of the team or group. Often this will be the CEO, Deputy CEO or Head of Department.

At the outset of a creative thinking training workshop, it can be intriguing to observe the discomfort of experienced healthcare professionals when announcing that, instead of having a traditional didactic training

The Working Format

•Meetings V •Workshops

- Inform by Exchange
- Instruct
- Defined Roles
- Many Issues

- Solve a problem
- Develop a plan
- Reach a consensus
- Usually focused on one
- Problem
- Promotes ideas and roles less defined

Uses Method to gather information and Engage participants in finding and Agreeing a solution

FIGURE 7.1 The working format.

session to strategize or problem-solve, they are going to engage in a workshop. Further, the workshop will use a gaming technique or play concepts. The introduction might begin:

> *OK everyone, we are going to have a workshop rather than a traditional meeting. Instead of going through actions and discussing agenda items we are going to play with LEGO®. This will get us better ideas and learning outcomes and we will get there faster!*

When using LEGO® SERIOUS PLAY®, we work to get the group of healthcare professionals to engage actively:

- Expressing their thoughts and ideas on the challenge using play through the metaphor of LEGO®.
- Modifying their ideas and sharing them with the other team members so as to inform the team of choices as they work towards a response to the challenge.
- Developing an understanding of the views of others using methods such as landscaping to create a shared model.
- Bringing their ideas together and identifying the connections among their individual models in the shared landscape model.
- Identifying the priorities across the landscape model and using them to develop a shared vision and plan of action.

In our experience, the LEGO® SERIOUS PLAY® methodology can engage healthcare professionals at all levels in the organisation. Beforehand, they may be so invested in "business as usual" and used to working largely in meetings with the same group that it is difficult for them to see a different path to deal with more strategic challenges. The LEGO® SERIOUS PLAY® methodology surprises and disarms them as it helps them to address problems and challenges as a team through legitimising a different form of expression. En route, they develop a deeper understanding of the problems and challenges, and empathy with their co-workers and the patients they serve. Overall, playing with LEGO® can make the individuals and the team more aware of the greater shared purpose they serve together, and how they can stay on the path to getting there as a team. It reminds them that, in a sense, they are all in the same boat with, ultimately, the same goals and challenges.

7.4 USING PLAY IN HEALTHCARE: DESIGNING THE ADAPTABLE PATIENT EN-SUITE IN A CHILDREN'S HOSPITAL

We end this chapter with a case example to illustrate the playfulness methodology and approach in practice. Creating an environment for creative thinking, design and innovation, we often bring teams of collaborating healthcare staff together in a workshop and ask them to prototype their ideas for delivery. One such challenge is to design a patient en-suite bathroom facility in a children's hospital.

7.4.1 The Background, the Problem and the Action: Designing an Adaptable Patient En-suite Bathroom Facility in a Children's Hospital

7.4.1.1 Background

The children's hospital is focused on providing child-centred care at a new facility under development at an existing hospital site. One of its key initiatives is to ensure that every child and their family has access to a private room if a stay is required. The new facility will not only provide this accommodation but also individual en-suite facilities. This worthwhile aim demonstrates how the team developing the overall hospital is putting the child patient first in designing systems to deliver all aspects of their care.

7.4.2 The Problem (Rather Than the Puzzle)

The children's hospital will cater for patients from very young infants (from a few weeks old) to young adults (up to18 years old). Obviously, an accommodation room and en-suite space will need a different configuration for different aged children. The task is to design the en-suite bathroom space to cater for such requirements as changing facilities for babies and infants, and bath facilities for able and disabled children and young people. General parameters must be considered such as compactness, cost, safety, infection control and potential use of technology.

7.4.3 The Action

Before collaborating in the design of the en-suite, the participants engaged in a LEGO® SERIOUS PLAY® workshop. This play-based workshop experience developed trust and confidence among the team members and created an environment for creative thinking, design and innovation. Then, when approaching the en-suite design brief, the team engaged in a creative process with strong emphasis on human-centred design and the user experience. In the event, they came up with the metaphor of the Hospital as exhibiting a Home, Village and City structure. In this setting, the patient should experience the room just like their home, or as a home from home. They should feel comfortable and safe, and they should be able to have family experiences there. The ward where the patient room is located should be like a village. Here, everyone knows each other and are friendly and supportive, helpful and caring, with all the home supports available. The wider hospital is the city. It should represent all that is good about a modern city with access to facilities, easy to get around, support for other needs such as entertainment, goods, food and play.

Articulating this metaphor, the team designed the en-suite to enhance the home-from-home aspiration for the room. They would not have felt enabled or empowered to think so creatively if the environment was not safe for them to play. Figure 7.2 shows one of the team members, a healthcare professional, demonstrating the team's ideas for solving the en-suite problem. In this case, rather than just using whiteboards or PowerPoint slides, the team used sticky notes, coloured card and paper. They commandeered a corridor outside the workshop room and, using paper, pens and a chair,

FIGURE 7.2 Design and demonstration.

they demonstrated and pitched their ideas to a wider team of stakeholders. They brought their concept to life in a simple but meaningful prototype which was accessible to all. The resulting discussion was rich and informed. They played the game!

7.5 CONCLUSION

The ultimate aim of any healthcare organisation is to provide an excellent quality healthcare experience for all its patients. At first glance, it might be tempting to ask what play, design and creativity have to do with delivering healthcare quality.

The concept of continuous improvement, now well accepted in the world of quality management, teaches us that every new solution will eventually become a problem that constrains achievement. As the problems and challenges in a healthcare organisation become increasingly complex and interconnected, and the effects of these interconnections become more prominent, the creativity required becomes more demanding.

For this reason, healthcare organisations need to become well-versed in creativity practices to be responsive, confident and structured enough to come up continuously with new ideas and methods to improve practice. In effect, they need to create a growth culture and environment where there is never a moment when they might say, "*everything is fine now, and we are providing the best quality of service that we can*". Rather, the challenge is more like "*how can we ideate, create and innovate to improve every day?*" And it is here that gaming has a role in creating a safe environment for creative thinking, design and innovation.

SOURCES AND SUGGESTED FURTHER READING

Caillois, R. 1961. *Man, play and games*, Trans. by M. Barash, Chicago: University of Illinois Press.

Cirella, S., Canterino, F., Guerci, M. & Shani, A.R. 2016. Organizational learning mechanisms and creative climate: Insights from an Italian fashion design company, *Creativity and Innovation Management*, 25 (2), 211–222.

Coughlan, P. & Coghlan, D. 2011. *Collaborative strategic improvement through network action learning: The path to sustainability*, Cheltenham: Edward Elgar.

Jobst, B., Köppen, E., Lindberg, T., Moritz, J., Rhinow, H. & Meinel, C. 2012. The Faith-Factor in Design Thinking: Creative Confidence Through Education at the Design Thinking Schools Potsdam and Stanford? in, H. Plattner et al. (editors.), *Design Thinking Research, Understanding Innovation*, Berlin Heidelberg, Springer-Verlag, pp. 35–46.

Kolb, A.Y. & Kolb, D.A. 2010. Learning to Play, Playing to Learn: A Case Study of a Ludic Learning Space, *Journal of Organizational Change Management*, 23, 26–50.

Ortony, A. (editor). 1993. *Metaphor and thought* (2nd edition), Cambridge: Cambridge University Press.

Powell, D.J. & Coughlan, P. 2020. Rethinking lean supplier development as a learning system, *International Journal of Operations & Production Management*, 40 (7/8), 921–943.

Schein, E.H. 2014. The role of coercive persuasion in education and learning: Subjugation or animation?, *Research in Organizational Change and Development*, 22, 1–23.

Schulz, K.P., Geithner, S., Woelfel, C. & Krzywinski, J. 2015.Toolkit-based modelling and serious play as means to foster creativity in innovation processes, *Creativity and Innovation Management*, 24 (2), 323–340.

Styhre, A. 2008. The element of play in innovation work: The case of new drug development, *Creativity and Innovation Management*, 17 (2), 136–146.

Wilson, F.R. 1999. *The hand: How its use shapes the brain, language, and human culture*, New York, Random House USA INC International Concepts.

Zenk, L., Hynek, N., Krawinkler, S. A., Peschl, M. F. & Schreder, G. (2021). Supporting innovation processes using material artefacts: Comparing the use of LEGO bricks and moderation cards as boundary objects, *Creativity and Innovation Management*, 30 (1), 845–859.

CHAPTER 8

Putting It All Together

How Might You Move Your Healthcare Setting towards an Innovation Culture and Mindset?

ONE OF THE AUTHORS was facilitating a series of workshops. He noted that the participants were gathering and ready for the third workshop:

> *I was working with the team finalising plans to take the group through a 3-hour journey to bring us to the next step. I noticed Alice seemed a little sad and worried. I asked her what was wrong. She replied "I like being here at the workshop but the ward where I work is short staffed. I feel very guilty that I get to be here, but while I am here my colleagues must work on and, in fact, must work harder because they have fewer staff without me". All I could say to her was thank you for coming and giving her time and that we would do everything to make sure the time she had spent with us was valuable.*

There is no question – innovation in healthcare is difficult, challenging and, at times, personal. In this book, we have been trying to address those difficulties. We have introduced key concepts to give you a framework and a language to enable you to question, take action, reflect and learn. We have described mechanisms and activities to develop your ideas, particularly in collaboration with different stakeholders. However, you might ask still "*so, tell me again, how can I do* this?" In response, we could have written a neat

DOI: 10.1201/9781003123910-8

summary chapter, revisiting the ideas introduced earlier. Instead, however, we bring these ideas together as we take you through a real example of an innovation initiative aimed at developing a new children's hospital in Ireland. The initiative has been a lived experience for the authors – one to a greater degree than the other. This work provided a proving ground and a demonstrator of the effectiveness of our ideas. In terms of Chapter 1, this initiative has been both an opportunity and a continuing challenge.

Supporting children's health requires many of the skills which are common in healthcare delivery and other segments of our society. However, children represent a particularly vulnerable group with health needs, often related to their young age. Many cannot articulate what they feel or want. Typically, children have the warmth and support of caring parents who accompany them on their healthcare journey. They translate, advocate and intervene as effective co-deliverers of healthcare. However, the parents also have their own concerns and needs for reassurance as their vulnerable children undertake this journey. So, the delivery of children's healthcare places particular demands on the staff and the facility where it all takes place. Of course, the facility must not only be a hospital, but also a playground for the children, and an accommodation facility for otherwise healthy parents. Said in terms of Chapter 4, this initiative requires an understanding of multifaceted problems and challenges and setting about solving them.

For many years, delivery of children's healthcare was spread over a number of smaller hospitals in Ireland. In 1993, a decision was taken to concentrate much of the delivery from three base hospitals in a purpose-built national hospital, focused entirely on children. The need and opportunity for innovation were recognised from the outset.

At the time of writing this hospital is under construction, a unique building forming one of the largest structural projects by the Irish Republic since its formation 100 years ago. Its design is in keeping with the novelty of the experience in prospect for the children, their parents and the staff who will visit, work and are cared for in it. However, the layout of the facility is to be central to the tasks undertaken by the healthcare delivery staff, the flows of patients through the many steps of their care, and the realisation of their duty of care. But how should this innovation be undertaken, who should be involved, how would they contribute and collaborate? These were some of the questions posed to and by the interdisciplinary team responsible for development and delivery of the hospital. In terms of Chapter 5, we challenged all to have the creative confidence to think differently, to empathise differently and to collaborate differently. In the following sections, we describe and reflect upon the evolution of these questions and responses. If, as a reader, you are beginning this chapter having engaged with the earlier chapters, you will recognise already the steps taken and methods employed. Different, however, will be the connected nature of the steps and method. In essence, they form a programme of collaborative innovation underpinned by what we call hybrid healthcare thinking.

8.1 THE SETTING

The setting for this initiative is 24 months into the building programme for the new hospital and 36 months before the planned opening. Three base hospitals were to merge into the larger entity that would be the new hospital. The innovation team was charged with innovating in a number of areas. The three members of the innovation team were (1) one of the authors – a scientist with considerable experience teaching and training in the innovation space and with practical experience related to invention and enterprise in healthcare. (2) An engineer with skills and experience developed in systems engineering and systems thinking related to the development and installation of healthcare information systems and their links and use with medical device technology. (3) A Senior Manager with a background in facilitation and organisational development, in particular, the area of people and change. However, these members were really the facilitators, and the actual innovation was carried out by the team members which included an interdisciplinary group of staff, parents and patients, all of whom were either, at the time, actively involved with the healthcare system or had recent experience of it. In terms of Chapter 3, the setting required the structure of the healthcare organisation to accommodate the healthcare professionals in generating pathways to deal with new ideas.

8.2 INITIATING HYBRID HEALTHCARE THINKING

The impetus of the new physical hospital facility helped to focus people on change and improvement. From the outset, the innovation team recognised the need to initiate an entirely new process for developing appropriate work practices for the new hospital development. They recognised that some starting principles were universally accepted within the team and, indeed, by the wider workforce in the three base hospitals. They grouped these principles into established, mythical and new working principles. At a later stage, the new working principles would need to be accepted by the staff, the children as patients who were already attending the base hospitals and their parents and families who supported them.

8.2.1 Starting with Established Principles

There were four established principles which reflected the current operation of the base hospitals:

- Everyone accepts that the current hospitals do not work as effectively as a system.
- Yet, most hospital employees and healthcare professionals carry out their work and duties to the best of their abilities on a day-in, day-out basis.
- Most hospital employees and healthcare professionals are dedicated and patient focused.
- Hospitals in the healthcare system have, for many years, been ineffective at utilising digital technologies to make their systems and processes more efficient and effective, this is falling way behind other business or government sectors.

The team recognised that, if not challenged, these established principles would constrain the innovation planned for the new hospital.

8.2.2 Mythical Principles

Each of the base hospitals, like other organisations, had built up a set of myths which embodied assumptions that would have to be debunked:

- When the new building opens, it will fix everything.
- Someone other than me is responsible for the things that go wrong outside of what I do to the best of my ability and within my professional remit.
- You need experts in the room to solve complex systematic healthcare problems.
- The senior executives and managers have the solutions to all our problems.
- A lack of resources is solely the cause of all of our problems.
- Digital technology and computers will come in and solve all our efficiency problems.

Unwittingly, these myths externalised current shortcomings and would place unreasonable expectations on the new hospital. If not debunked, they could hamper realisation of the promise of the new hospital and lead to a new level of resistance to change.

8.2.2.1 New Working Principles (in Evidence) to Be Adopted

Finally, a set of new working principles was identified which, if enacted, could lead to realisation of the potential of the new hospital. At their roots, these new principles could address shortcomings endemic in the current system:

- The problems and challenges that are faced in the efficient and effective delivery of healthcare cannot and should not be pigeonholed into the category, silo or type of activity we think we need to address i.e., clinical, operations, human resources or process. Rather, they should be seen as multifaceted and therefore requiring a multifaceted, interdisciplinary approach to finding a solution and solving the problem.
- Even if a problem or challenge seems to relate to just one activity in healthcare, finding a disruptive, creative and even innovative solution to the problem will require bringing together a wide range of bright, engaged people to try and solve it. Having a group of people of the same age or with the same professional backgrounds will reduce the chances of such a solution emerging.
- You do not need to be an expert in the problem principles to be involved in finding a creative solution.
- Problems we face can be better, more efficiently and effectively addressed by using facilitated innovation workshop techniques, rather than how we have traditionally tackled them using sequential and often unstructured meetings.
- A broad and inclusive range of colleagues in the organization can contribute professionally and creatively to solving the multifaceted problems that we face.
- The three cornerstones we can use to effect change in the problems and challenges we face in healthcare relate to people, process and technology. We need to accept that technology is only a tool for this transition, and we need to make effective changes to ways of working and systems thinking, which relate more to people and process if we

are to make meaningful, effective, and lasting change in the healthcare system.

These new principles are recognisable in terms of Chapters 5 and 6.

8.3 HYBRID HEALTHCARE THINKING

The innovation team's first project was related to services being developed at the new hospital. The challenge was to imagine how the new wards were going to work in a facility that was still under construction. The innovation team was pleased when the Director of Nursing for the new hospital took on the role of executive sponsor for the project. She was a member of the executive management team and, as a sponsor, was to play a critical role in getting access to the people, facilities and materials we needed to run a series of workshops. In particular, she provided the communications conduit between senior management and the innovation team. This conduit was essential as, in the early days, the senior management group was somewhat sceptical of the approach in prospect.

Through a series of meetings, the innovation team designed a workshop plan to implement the new thinking in action and, so, respond to the challenge. We refer here to this new workshopping structure as "*Hybrid Healthcare Thinking*", illustrated in Figure 8.1. With the support of the project sponsor, the plan was approved by the executive team. As was necessary, a project initiation document (PID) was developed. In this phase, the innovation team was enacting Chapter 7 as they looked to create an environment for creative thinking, design and innovation. So, from the outset, they realised that, when implementing the plan, it was important that those involved:

- Worked as an interdisciplinary team.
- Worked in an environment where everyone could develop and demonstrate their creative confidence and, so, were unafraid to explore and express the ideas they had, or would have, during the workshops.
- Would step out of their comfort zones and see how collectively they could be involved in co-designing the new ways of working and optimising for the new hospital.
- Understand the facilitation and the innovation methods needed, and to be open to developing their individual skills in that regard.

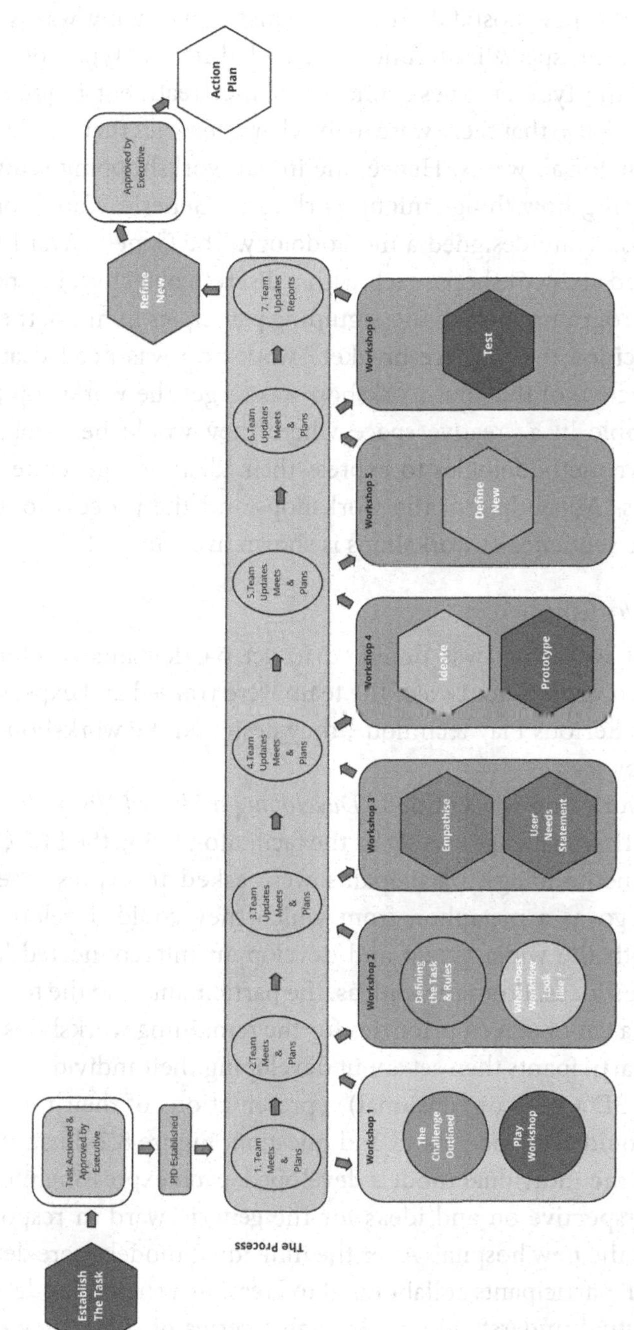

FIGURE 8.1 Hybrid healthcare thinking.

The innovation team began with an exploration of how wards would work in the new hospital. They recognised that many wards might have particular or specialised functions, and different types of ward would relate to the type of illness suffered or the treatment required. Yet, they recognised also that there were many characteristics that would be the same or similar for all wards. Hence, the initial workshopping sequence began by exploring how things might work in a "Generic Ward" for which the innovation team designed a methodology. The Generic Ward Programme comprised six workshops, each with a discrete objective, but related to the overall programme. With these guiding principles in mind, the innovation team decided that an "Ice-breaker" workshop was needed at the outset. The objective of this first workshop was to get the workshop participants comfortable in a creative space where they would be using playfulness and other methodologies to express their ideas and generate solutions to problems. An outline of the workshops and the process to approve and enact the sequence of workshops is shown in Figure 8.1.

8.3.1 Workshop 1

This first workshop was designed to get participants comfortable using play techniques. Since two of the team were trained and experienced using the Lego Serious Play technique, they designed the workshop around this methodology.

The workshop was entitled "*Developing a Shared Vision for the Generic Wards*". This challenge was set by the facilitator. Using the LEGO® SERIOUS PLAY® methodology, participants were asked to express their thoughts using Lego as a metaphor, from which they could develop ideas, share them with the wider group and develop an interconnected landscape of everyone's ideas. From these ideas, the participants and the team would co-develop a list of agreed priorities for the remaining workshops.

The participants then set about developing their individual Lego models creating 3D (three dimensional) representations of their thoughts so that others could see, understand and question. Figure 8.2 illustrates the landscape of the individual models developed, each expressing metaphorically their perspective on and ideas for the generic ward in response to their needs in the new hospital. After the individual models were developed and explored, participants collaborated in creating a shared model to represent their mutual understanding. Through a series of steps, they explored the interlinkages and connections between their individual and shared ideas and, ultimately, set out an agreed list of priorities.

FIGURE 8.2 The landscape of the individual models.

On reflection, the innovation team members were satisfied with the outcome, identifying a common purpose. From this there was clear evidence that a group of creative and innovative participants could work together and engage effectively in Hybrid Healthcare Thinking. The outcome represented new actionable knowledge for the innovation team.

8.3.2 Workshop 2

8.3.2.1 Defining the Task and the Timeline

Having formed as a group in Workshop 1 and agreed on priorities, the next step for the participants was to define the tasks and timeline for exploring how the new wards were going to work in practice. Workshop 2 required a variation from the traditional steps for a design thinking style workshop. The innovation team believed that, due to the enormity and complexity of the overall task, it was important that the participants developed a shared sense of what it was about. This was as much to inform them of the scale and complexity of the task, as it was to motivate them to be creative in generating ideas as to how the various problems and challenges could be solved. Developing a shared understanding of the problem as a group at this stage was paramount.

Stakeholders such as the patients, their families and the ward staff were identified. A visualization of the process that each stakeholder would go through helped the participants to understand the needs and experience of each one. The innovation team used a technique known as basic journey mapping to set the scene and define the scope of what became known as the service blueprint for the generic ward. The innovation team member familiar with systems thinking generated some ward-related material from the layout drawings for the new hospital. It is worth pointing out that, while this journey mapping process was all about designing new ways of working on the ward, one of the major constraints was the physical building. This building, although new, had been through a separate design process and was well on its way to being constructed. So, changing the building was not an option. However, this workshop was a chance for the participants, as representatives of the future hospital's staff, to review and revise decision-making rules, constraints and enablers to accommodate the service blueprint in the new building. The workshop also was to establish the task and timeline for this innovative work.

The innovation team member leading the workshop established a simple method for looking at ward flows and narratives for how various patient care activities might be carried out. So, when he described the layout of the new ward, he started with the architect's drawing of the ward layout (Figure 8.3).

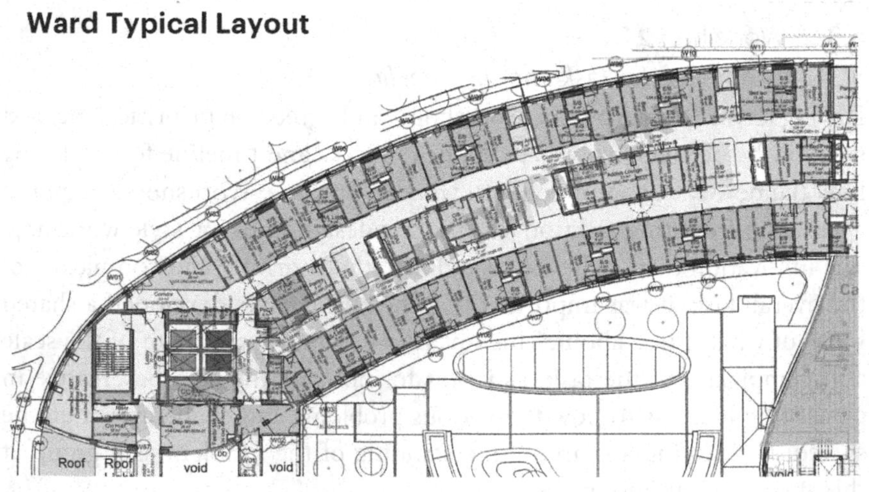

FIGURE 8.3 Typical ward layout.

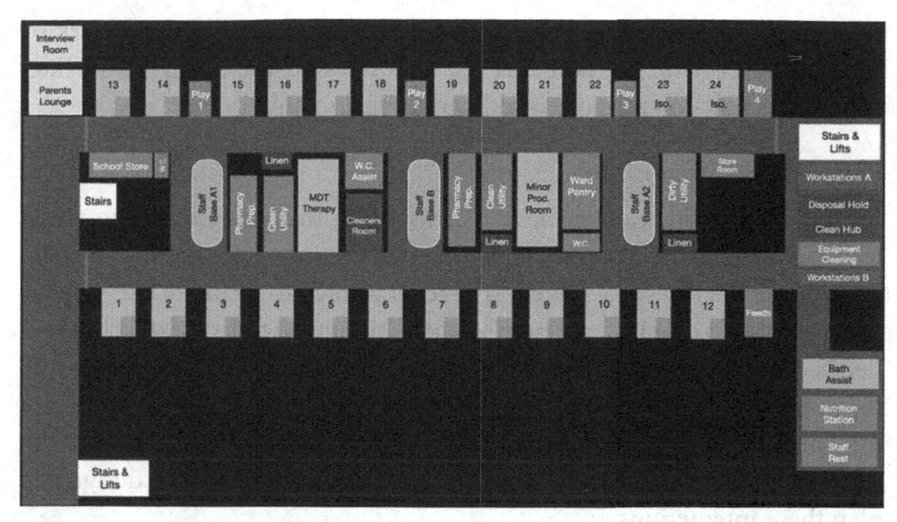

FIGURE 8.4 Interpreting the layout.

As the building was curved, the drawing was also curved. As such, the innovation team could see that it would be difficult for busy healthcare staff to relate to this curved space and decided to simplify it visually. The team used prototype thinking and produced a more visually pleasing interpretation of the layout, shown in Figure 8.4.

In the event, the workshop participants connected with this interpretation immediately and it helped them to imagine realistic scenarios related to day-to-day activities in the ward. In essence, the innovation team was using this simplified 2D representation of the ward as a metaphor to help the workshop participants to visualise and experience at some level what it would be like to work on the new ward. Their mindset was so in tune with the challenge that the participants brought back copies of this ward layout and displayed them proudly in their existing work areas and offices. Recognising the effectiveness of this visual device, the innovation team used the diagram in many scenarios at later workshops.

The main output from Workshop 2 was that participants had a real sense of the novel and what was possible. If the innovation team had not taken this approach, many participants would have had only their current work environment as a point of reference. In such cases where staff were working in older, unfit for purpose facilities, reimagining the ward layout would have been extremely difficult for them. In contrast, Workshop 2 enabled them to recognise the building constraints, re-imagine the new and plan to co-design in an interdisciplinary way. In reality, the participants were working as co-designers. They could ideate their new ways of working together to provide the best care for children and their families in their new modern fit-for-purpose healthcare environment.

8.3.3 Preparation for Workshop 3

As the innovation team planned Workshop 2, they were thinking ahead to Workshop 3. In Workshop 3, they would focus on identifying and confirming the key stakeholders. The team wanted to facilitate participants on a pathway to developing a more empathetic approach to the stakeholders' needs. Consistent with Chapters 4 and 5, the innovation team wanted to focus on the user/stakeholder experience, and to have some simple but informative preliminary interactions with key stakeholders. So, the team organised a short session (30 mins) at the end of Workshop 2 to design and plan these interactions.

Empathy is very much related to the user experience. So, recognising the children and their parents as the main stakeholders, the preliminary enquiry was based on some straightforward empathetic questions about their feelings before and during their time on the ward. Table 8.1 outlines the questions.

The discussions with the Workshop 2 participants helped the innovation team to determine that there were four stakeholder groups to be focused on and met. Correspondingly, a plan of action was developed, as summarised in Table 8.2.

8.3.4 Workshop 3

Based upon the preparatory work, Workshop 3 focused on the broad theme of empathy as an important step in the design thinking methodology. It would be used as a tool to design with and also to acquire insight into users'

TABLE 8.1 Empathy Questionnaire for Users

Question	Content
1	Before coming to the Ward what did you hear (e.g., from friends, family, healthcare provider, internet, media)?
2	What were your thoughts and how did you feel before coming into the Ward?
3	What was your experience of coming into the Ward?
4	What did you see on the Ward, what are the things that stood out for you?
5	What were the pains and challenges you experienced coming onto the ward?
6	What were the gains and opportunities for you coming onto the ward?

TABLE 8.2 Plan of Action

Stakeholder	Representative	Action
Patients/Children	Children on the Ward	3 from team meet and interview children
Parents	Parents on the Ward	3 from team meet and interview parents
Healthcare Workers	Other Members of Generic Ward Team	3 from team meet and interview healthcare workers
Executive Team	Chief Executive Officer	2 from team meet and interview CEO

needs. So, the conversation was to be directed by the data that emerged from the user experience interviews.

8.3.4.1 Empathise

The workshop introduced design thinking as an innovation tool of choice to foster innovation and creativity in organizations as they attempt to address complex problems. It was re-emphasised that, in a design thinking workshop series, all participants in the interdisciplinary team needed to be empathic with the users they were designing for in order to better create relevant solutions. The innovation team wanted to facilitate the participants to see empathy as having two main dimensions: emotional and cognitive. Emotional empathy was based on shared and mirrored experience where a person felt what others experienced. In contrast, cognitive empathy was based upon an understanding of how others might experience the world from their point of view. In particular, the innovation team wanted to use simple human design principles, as described in Chapter 5, to develop and refine the problem statement about how the wards in the new hospital would work.

The plan was to take the participants through a shared experience that would:

1. Immerse them in empathy for the key stakeholders identified earlier and develop a set of empathy maps.
2. Develop a journey map.
3. Refine and redefine a set of user needs statements for each identified key stakeholder.

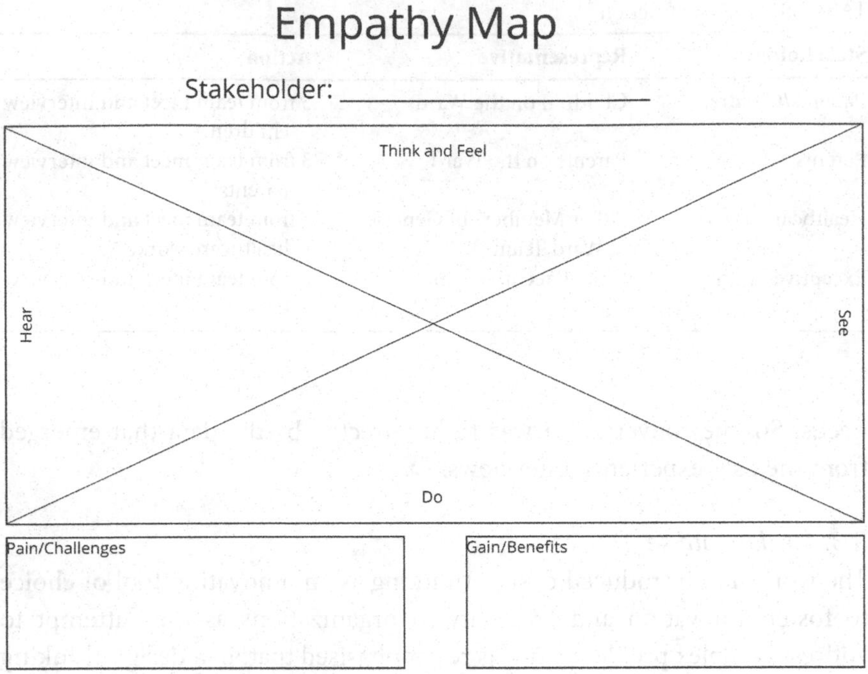

Empathy Map

Stakeholder: ______________________

Think and Feel

Hear

See

Do

Pain/Challenges

Gain/Benefits

FIGURE 8.5 Template for an empathy map.

We provide templates and examples in Figures 8.5–8.9.

First, the innovation team introduced the simple template above for an empathy map (Figure 8.5) and set about populating the map for each of the stakeholders. They broke the participants into four groups. Each group would focus on one of the key stakeholders and complete a map.

When completed, each map was visible to all, visual and easily interpreted. The discussion was rich, evidence-based and informed by the user experience interviews. The empathy map for Ward Staff as stakeholders and healthcare workers is shown in Figure 8.6.

On completion of the Empathy Maps, the innovation team facilitated a discussion and asked the participants to complete Journey Maps, using the template in Figure 8.7.

An example of how the participants used the journey map template is shown in Figure 8.8.

Finally, the workshop finished with the development of the user needs statement, using the template in Figure 8.9.

At the end of this process, the participants developed a set of user needs statements for the four identified stakeholders:

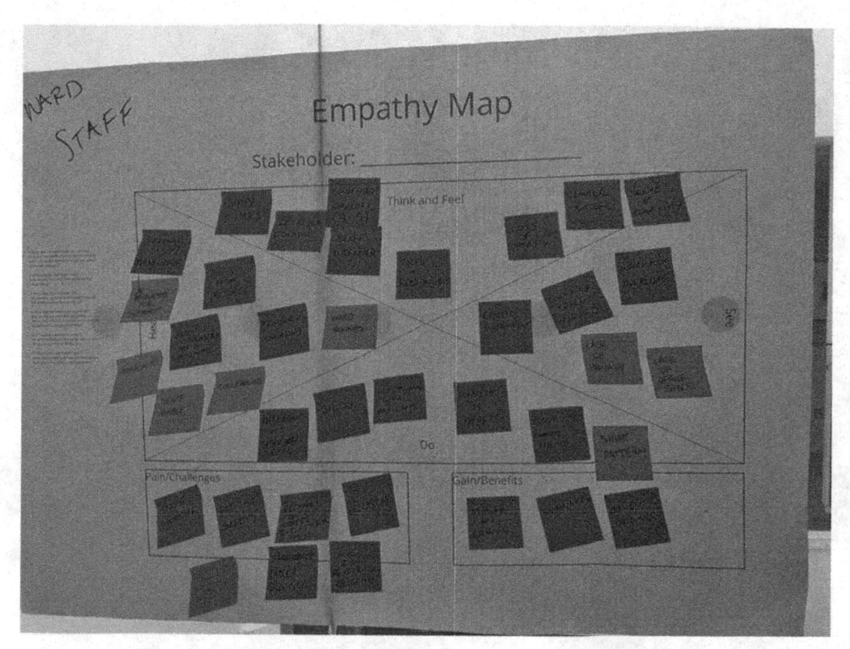

FIGURE 8.6 Empathy map for ward staff.

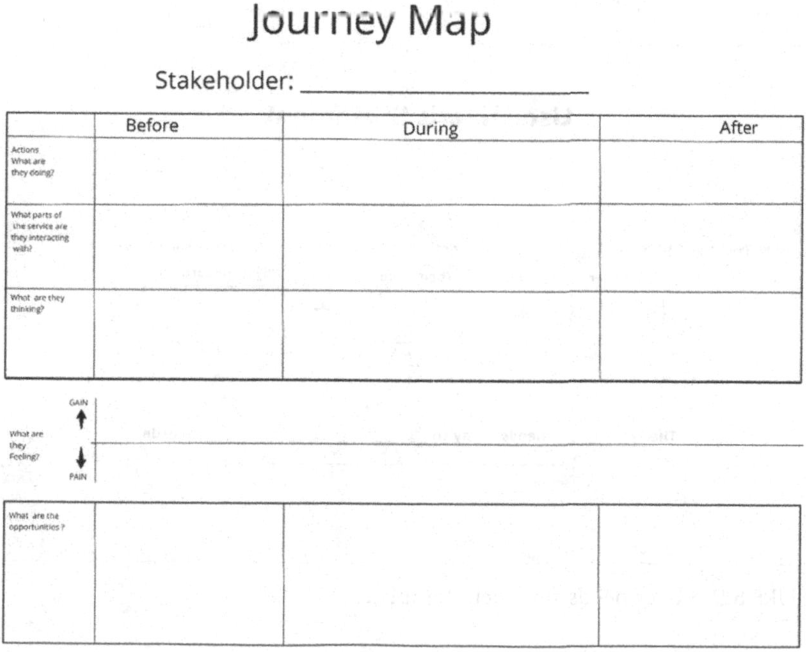

Journey Map

Stakeholder: ______________________

	Before	During	After
Actions What are they doing?			
What parts of the service are they interacting with?			
What are they thinking?			

What are they Feeling?	GAIN ↑ / ↓ PAIN

What are the opportunities?			

FIGURE 8.7 Journey map template.

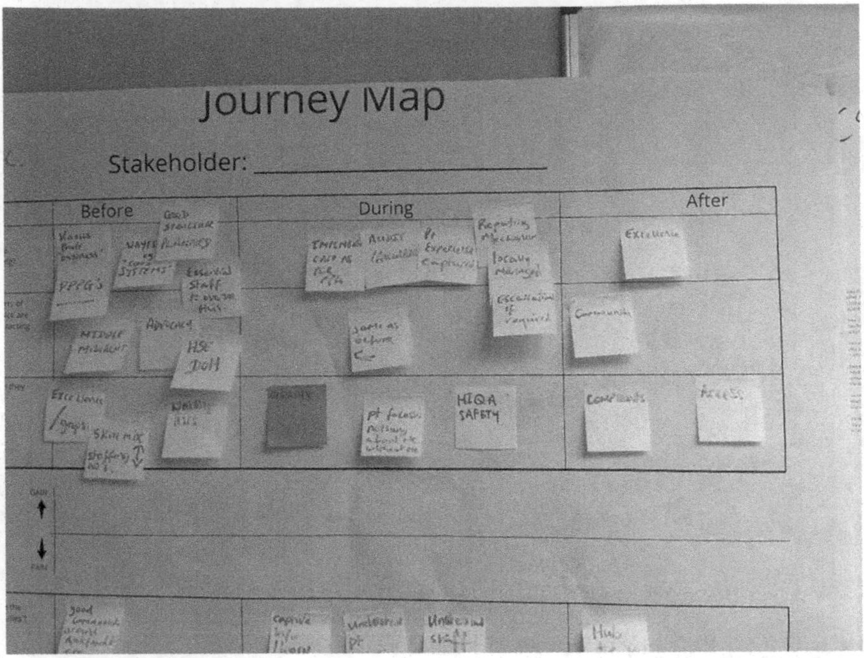

Journey Map

Stakeholder: ______________________

Before | During | After

HIQA SAFETY

FIGURE 8.8 Sample journey map.

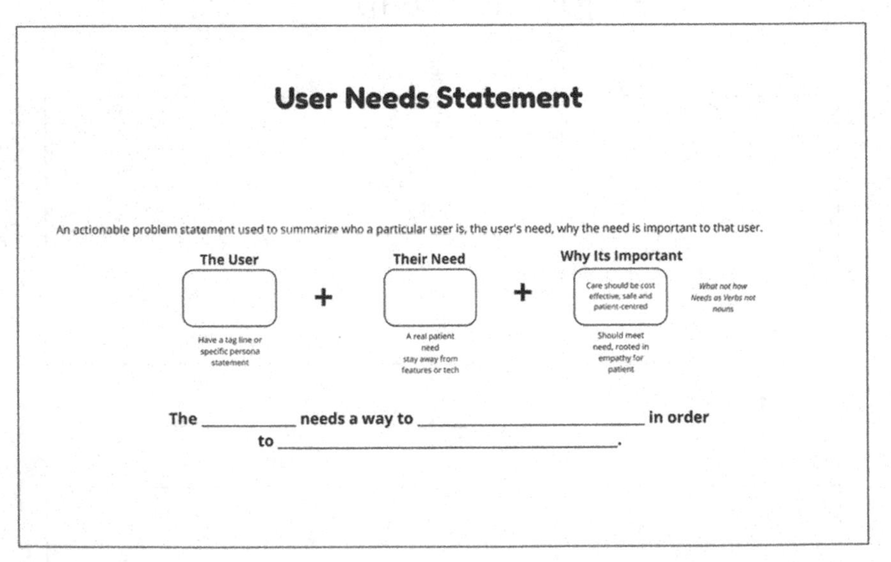

FIGURE 8.9 User needs statement template.

1. *The CEO and Management need a way to ensure the best patient and family experience and in order to deliver safe world-class clinical care that is child centred, compassionate and progressive.*
2. *The Ward Staff need to feel supported and appreciated to provide the most positive outcome for patient/parent.*
3. *The Child and Young Adult need a way to be involved in their care to feel safe, respected and heard.*
4. *The Parent/Guardian needs a way to access expert care (for their child) to achieve optimum personalised outcomes.*

These statements redefined the problem in terms of the stakeholders and gave the participants a more empathetic approach to thinking creatively and coming up with new ideas for solving the challenge.

8.3.5 Workshop 4 – Ideate and Prototype

Now that the participants had a very explicit user-defined statement for the four main stakeholders, Workshop 4 was oriented to look at methods to generate ideas for solving the problems. The aim was to think creatively about future ways of working for the generic ward, to generate new ideas/ innovations, as well as consider what to amplify and what to reduce in the current practice. The innovation team and the participants agreed that this aim was best approached in two parts:

A. What changes can we make, and what innovations can we adopt to help us reach our purpose?
B. What do we reduce and what do we amplify?

To carry out Part A, the innovation team invited participants to ideate on their own. They were also encouraged to use whatever materials they needed. Many made collages and drawings on the walls. Others used paper and post-it notes. Figure 8.10 is a notable example of how one team member used a boat analogy to show the importance of communication on the ward in order for patients to find their way around. The innovation team then encouraged the participants to combine and organize their ideas and thoughts and to agree or poll on the best ideas.

As a second step in Part B, the innovation team asked the participants to consider what they might reduce and what to amplify. Again, the

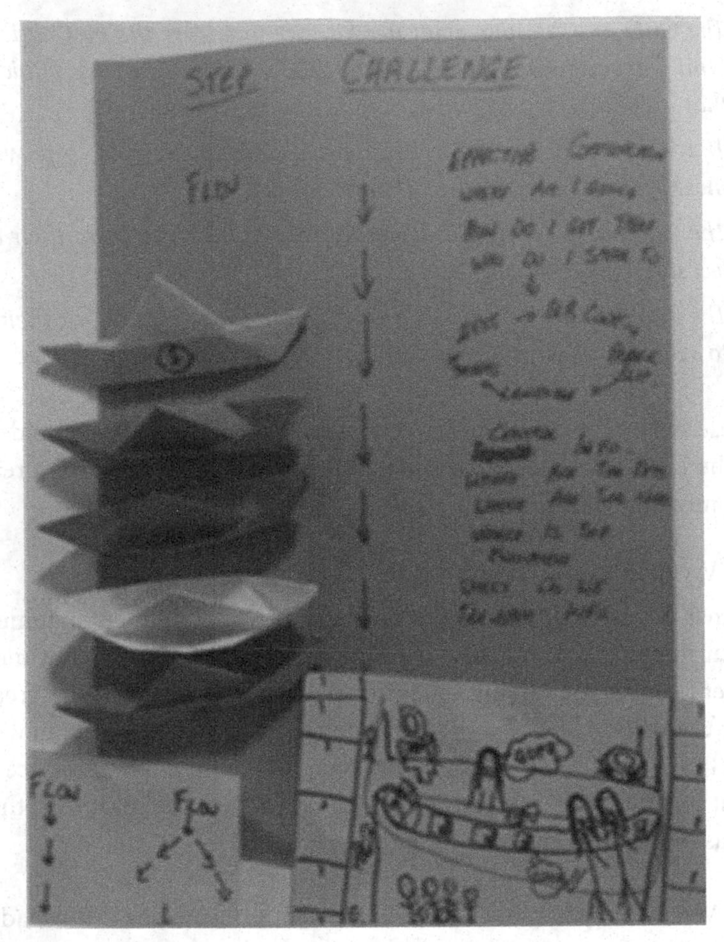

FIGURE 8.10 Showing the importance of communication on the ward.

participants were facilitated to ideate on their own initially. The focus was on what they could reduce in terms of current activities and also what they should amplify in terms of what was good. Then they combined their ideas and organized the emerging set.

The outputs of Workshop 4 comprised a refined list of ideas, identifying what to amplify and what to reduce. Examples of some of the ideas are in Table 8.3.

At this stage, one of the overriding themes of the ideation process emerged: the patient's room on the ward should be seen as their home from home, the ward itself should be a village and the wider hospital was the city and represented all that was good about the city (Figure 8.11).

TABLE 8.3 A Refined List of Ideas

Ideate Workshop

CITY – VILLAGE – HOME WARD IS WORKING FOR EVERYONE ELSE – EVERYONE ELSE SHOULD BE WORKING FOR THE WARD

- A mission Statement
- Quiet place for staff
- Changing room, lockers
- Parking space with name
- Healthy food, staff, patients and parents
- Stronger replenishment/replenishment at night
- A contact index–who, where and number
- Wi-Fi
- Virtual tours for families of new build
- Easy to use tech
- Trained on new systems before opening
- Comfy spaces
- Teenage rooms
- Different colour blankets
- Step-by-step guide – not everyone knows the steps
- Communication – a text to tell parent/child where to go in new build
- Private rooms for communication with staff and families
- Storage, storage and storage – for staff, families and patients
- A waiting area on ward for discharge
- Journey map for families
- A meet and greet at main door – even out of hours
- Bleep system – not always sure if person has seen it
- Ideal world – if something is requested its done
- Meet household and plan day/night
- Pharmacy – amazon solution – all to do with flow
- Translator – English is not always the first language
- Feed the parent – helps their wellbeing
- Hotel shower kits for families – toothbrush, shower gel
- Ward is inclusive and welcoming
- Play therapist – enough of them to meet all children
- A patient passport – tells staff their meds, needs like physio
- Charts – HCR big improvements
- Everything has a place – a tag system for beds, hoist, equipment on ward stays with ward
- Allocated space is used for its purpose – does not become a storage room
- Celebrate staff – achievements, new roles, new positions
- Patient is in on the ward there would be a schedule plan – real fear of missing doctor
- Volunteers – huge
- Left hand friendly
- Being Green and Ecco-friendly
- Breaks and lunches – how long to walk to shop, café considering the side of build
- Roles re-imagined
- 24/7 support
- Working day is not 8–4

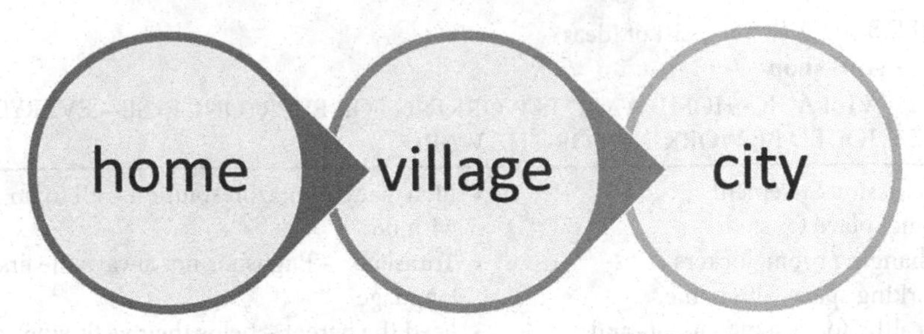

FIGURE 8.11 Emerging perceptions of room, ward and hospital.

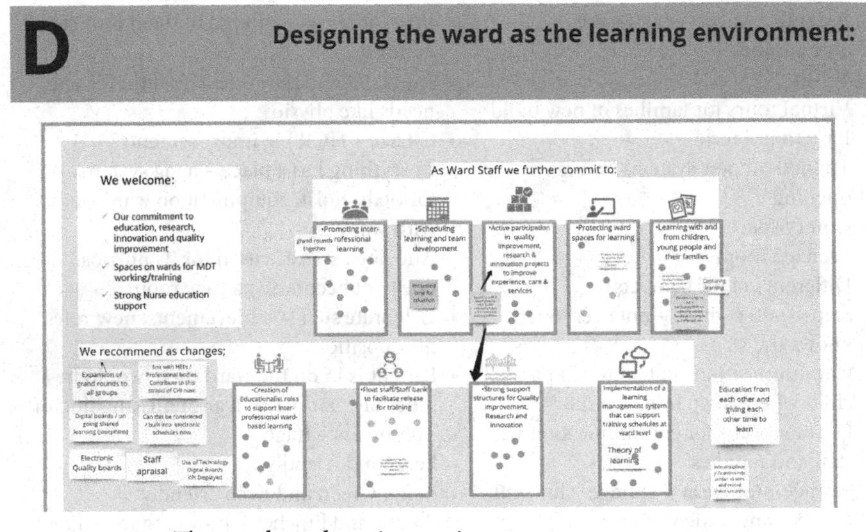

FIGURE 8.12 The ward as a learning environment.

8.3.6 Workshop 5 – Define the New

Workshop 5 built on the ideation from Workshop 4 and merged it into prototyping to demonstrate feedback between prototype and ideation. Ultimately, the thoughts and ideas of the participants could be broken into four major themes:

1. The Home/Village/City experience
2. The ward as the hub for clinical care
3. The efficient, well-organised ward space
4. The ward as a learning environment

The prototype laid out in more detail what these themes might look like. Figure 8.12 outlines details of the theme of the ward as a learning environment.

8.3.7 Workshop 6 – Test

After reflection of the series of workshops to date, the innovation team decided that the best way to test the emerging ideas was to present them back to the stakeholders in the form of an interactive workshop. As four themes had been highlighted in Workshop 5, the innovation team decided that these themes would be the focal points for Workshop 6. To test the plan, the team brought together representatives from the key stakeholders of the hospital development:

1. Executive management
2. Ward workers
3. Children
4. Parents and guardians

This work was in train during the Covid-19 pandemic in 2021 when a national requirement to restrict movements was announced just before the workshop. Working in an agile fashion, the innovation team quickly moved to convert and deliver the interactive workshop online.

Typically, in other workshops, a team would present or pitch their ideas to the stakeholders. However, considering the complexity of this project, the investment of staff time and energy into the process, the innovation team decided that workshopping some of the concepts together with the stakeholders was the most suitable forum. Then the stakeholders could comment on what was good and what could be improved on among the ideas and concepts presented. Figure 8.13 outlines the method which was based upon the use of a sophisticated online collaborative whiteboard system known as Miro Boards (Realtimeboard Inc., San Francisco, USA)

Figure 8.14 shows the actual online whiteboard that was used to run the online workshop. The four pillars from the ideas developed during the earlier workshops are shown as A, B, C and D. The workshopping participants and representatives from the stakeholders then visited each pillar where the participants outlined the ideas presented and invited the stakeholders to comment and advise.

Working in groups:

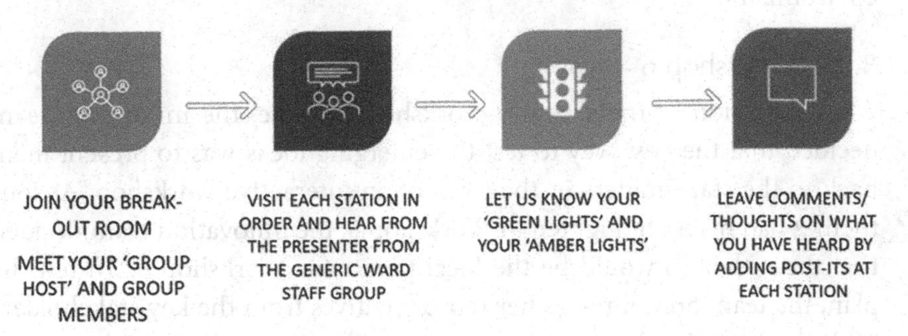
JOIN YOUR BREAK-OUT ROOM
MEET YOUR 'GROUP HOST' AND GROUP MEMBERS
VISIT EACH STATION IN ORDER AND HEAR FROM THE PRESENTER FROM THE GENERIC WARD STAFF GROUP
LET US KNOW YOUR 'GREEN LIGHTS' AND YOUR 'AMBER LIGHTS'
LEAVE COMMENTS/ THOUGHTS ON WHAT YOU HAVE HEARD BY ADDING POST-ITS AT EACH STATION

FIGURE 8.13 Working in groups.

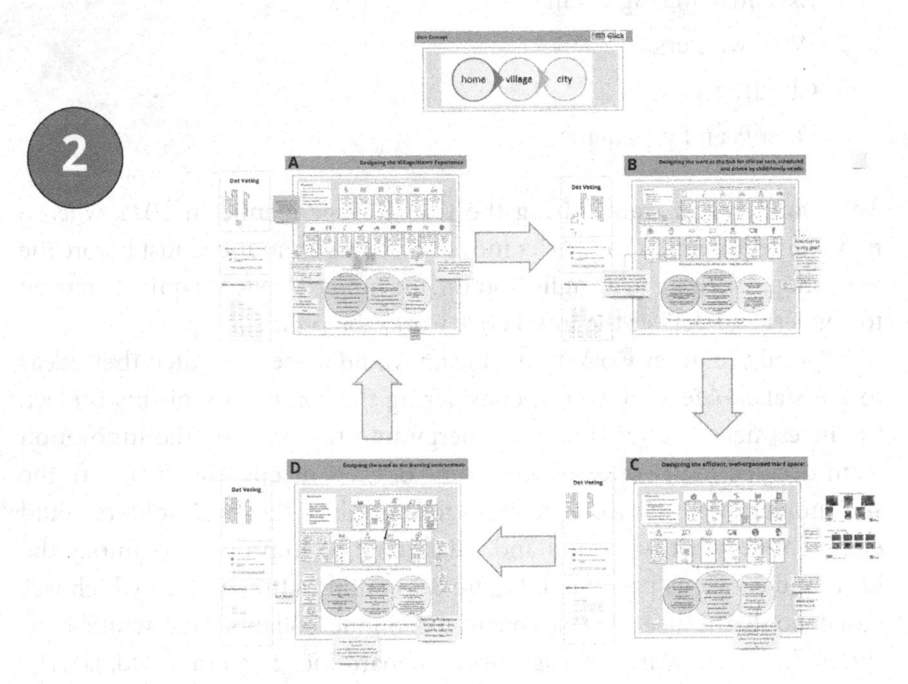
home
village
city
2
A
B
C
D

FIGURE 8.14 Group working using Miro Boards.

8.4 REFLECTION AND CONCLUSION

This series of workshops illustrates Hybrid Healthcare Thinking in action. It has delivered new insights and value for the innovation team engaged in the development of the children's hospital. The takeaways include:

- The importance of being flexible and able to pivot.
- The power of teamworking in a workshop context.
- The need to socialise the participants so that they can explain and pitch the emerging ideas to the actual stakeholders.
- The importance of getting the group into a physical and mental space where they could feel safe to be creative, ideate and ultimately be facilitated to produce disruptive solutions to the multifaceted problems they faced.

For the purposes of this book, the series marks a consolidation in practice of the ideas, tools and techniques introduced throughout the work and titled Hybrid Healthcare Thinking. The overriding message is clear: these tools and techniques are valuable individually. However, when used collectively and collaboratively, they blend to challenge the healthcare professionals to develop their thinking progressively, to act with confidence and to inject creativity and innovation into their improvement activities.

Index

Note: Page numbers in **bold** refer to tables and those in *italic* refer to figures.

Printed in the United States
by Baker & Taylor Publisher Services

Printed in the United States
by Baker & Taylor Publisher Services